SOCIAL RESPONSIBILITY

Failure Mode Effects
and Analysis

Industrial Innovation Series

Series Editor

Adedeji B. Badiru

Department of Systems and Engineering Management
Air Force Institute of Technology (AFIT) – Dayton, Ohio

PUBLISHED TITLES

Computational Economic Analysis for Engineering and Industry
Adedeji B. Badiru & Olufemi A. Omitaomu

Conveyors: Applications, Selection, and Integration
Patrick M. McGuire

Global Engineering: Design, Decision Making, and Communication
Carlos Acosta, V. Jorge Leon, Charles Conrad, and Cesar O. Malave

Handbook of Industrial and Systems Engineering
Adedeji B. Badiru

Handbook of Military Industrial Engineering
Adedeji B.Badiru & Marlin U. Thomas

Knowledge Discovery from Sensor Data
Auroop R. Ganguly, João Gama, Olufemi A. Omitaomu, Mohamed Medhat Gaber,
and Ranga Raju Vatsavai

Industrial Project Management: Concepts, Tools, and Techniques
Adedeji B. Badiru, Abidemi Badiru, and Adetokunboh Badiru

Inventory Management: Non-Classical Views
Mohamad Y. Jaber

Social Responsibility: Failure Mode Effects and Analysis
Holly Alison Duckworth & Rosemond Ann Moore

STEP Project Management: Guide for Science, Technology, and Engineering Projects
Adedeji B. Badiru

Systems Thinking: Coping with 21st Century Problems
John Turner Boardman & Brian J. Sauser

Techonomics: The Theory of Industrial Evolution
H. Lee Martin

Triple C Model of Project Management: Communication, Cooperation, Coordination
Adedeji B. Badiru

FORTHCOMING TITLES

Essentials of Engineering Leadership and Innovation
Pamela McCauley-Bush & Lesia L. Crumpton-Young

Handbook of Industrial Engineering Calculations and Practice
Adedeji B. Badiru & Olufemi A. Omitaomu

Industrial Control Systems: Mathematical and Statistical Models and Techniques
Adedeji B. Badiru, Oye Ibidapo-Obe, & Babatunde J. Ayeni

Innovations of Kansei Engineering
Mitsuo Nagamachi & Anitawani Mohd Lokmanr

Kansei/Affective Engineering
Mitsuo Nagamachi

Kansei Engineering - 2 volume set
Mitsuo Nagamachi

Learning Curves: Theory, Models, and Applications
Mohamad Y. Jaber

Modern Construction: Productive and Lean Practices
Lincoln Harding Forbes

Project Management: Systems, Principles, and Applications
Adedeji B. Badiru

Research Project Management
Adedeji B. Badiru

Statistical Techniques for Project Control
Adedeji B. Badiru

Technology Transfer and Commercialization of Environmental Remediation Technology
Mark N. Goltz

SOCIAL RESPONSIBILITY

Failure Mode Effects and Analysis

Holly Alison Duckworth
Rosemond Ann Moore

CRC Press
Taylor & Francis Group
Boca Raton London New York

CRC Press is an imprint of the
Taylor & Francis Group, an **informa** business

CRC Press
Taylor & Francis Group
6000 Broken Sound Parkway NW, Suite 300
Boca Raton, FL 33487-2742

© 2010 by Taylor and Francis Group, LLC
CRC Press is an imprint of Taylor & Francis Group, an Informa business

No claim to original U.S. Government works

Printed in the United States of America on acid-free paper
10 9 8 7 6 5 4 3 2 1

International Standard Book Number: 978-1-4398-0372-1 (Hardback)

Library of Congress Cataloging-in-Publication Data

Duckworth, Holly Alison, 1963-
    Social responsibility : failure mode effects and analysis / Holly Alison Duckworth, Rosemond Ann Moore.
        p. cm. -- (Industrial innovation series)
    "A CRC title."
    Includes bibliographical references and index.
    ISBN 978-1-4398-0372-1 (hardcover : alk. paper)
    1. Social responsibility of business. I. Moore, Rosemond Ann. II. Title. III. Series.

HD60.D83 2010
658.4'08--dc22
                                                                        2009047376

Visit the Taylor & Francis Web site at
http://www.taylorandfrancis.com

and the CRC Press Web site at
http://www.crcpress.com

We dedicate this book to the influence of W. Edwards Deming.

It is not enough to do your best; you must know what to do, and then do your best.

—*W. Edwards Deming*

Contents

Preface

Holly and Roz are both old-school manufacturing types. We've both spent lots of time on the shop floor in heavy manufacturing industries. We have engineering backgrounds. We love hands-on action-oriented shop floor work. Our point of view is not born from academia or politics or philosophy. It comes from seeing what can happen when "the pedal hits the metal" on the front lines of organizations. Some may see people with our background, writing about a topic of such seemingly untraditional views as social responsibility, as unusual. We disagree. We are also parents.

From our experience with continuous improvement (we are both Six Sigma Master Black Belts) we have also personally witnessed the waste, which can be so easily removed from the workplace. We are confident that most of our readers will also have firsthand experiences in this regard. So many of us see ridiculous situations in our organizations, situations that waste energy, waste resources, waste human effort, waste human potential, waste food—waste water, waste. This waste, and the ready acceptance of its existence, is actually material we are stealing from our children. We cannot continue to waste (for example, Americans reportedly throw out 200,000 tons of edible food daily) and expect the next generations to have access to the same standard of living. We must begin to create sustainable social structures for the protection of our children's health and well-being. This is our motivation.

We do not see social responsibility as a political movement. Our approach to social responsibility is diametrically opposed to a "socialistic" philosophy. We see social responsibility as a science, engineering, and business management imperative. We contend that scientists, engineers, and organizational leaders who fail to plan for the seventh generation in their product, process, and organizational designs are simply unskilled. Competence in these sciences, by our definition, means that we do not sacrifice performance—we do not sacrifice comfort, features, or growth while building systems that can be ecologically and socially sustained. It's not *either/or* but *and*. Is it more difficult? Sure. Does it take more careful planning? You bet. Does it require the use of specially designed tools to facilitate a balance of the complex interactions and requirements? Yes. We hope to present you with one of these tools in the following pages.

In our experience, we have learned so much about this *and* in the quality movement. As young, green engineers we entered the workforce on the cusp of the culture change caused by the quality movement. Our parents saw the need to trade cost for quality. You sought either a quality product or a cheap product. From the great teachers of quality, we learned and applied engineering and management principles that demonstrated that we didn't need to choose quality or cost. In fact, in so many professional endeavors, we found that the more we improved quality, the more profitable our organizations became, and a lower cost could be experienced by the customer. There is an economic benefit to a focus on product and process quality. The customer can have low cost *and* quality. In fact, it is our experience that low cost can only be sustained through a focus on the elimination of waste and defects.

It is this experience that has led us to the application of similar tools to the impending social responsibility movement. We don't have to choose either ecologically sustainable or inexpensive. We *know* both can be achieved simultaneously. We don't have to choose human-trafficked labor or expensive lawn care. We don't have to choose good corporate governance or profits. We don't have to choose fair labor practices or good customer service. We can do both. And there will be an economic benefit to the organizations that do both.

Currently, unethical practices create market advantage through the exploitation of unequal social responsibility standards throughout the world. Child labor in India creates inexpensive home decorating merchandise for the mini-mansion in Dallas, Texas. The shopper in Dallas is unaware of the exploitation that has created the item. The mother in India is reliant upon her children's labor to feed the family. Until we find solutions on both sides of this equation, the exploitation will continue. The answer is not to stop the child labor without recourse, nor is it to raise the price of home decorating items. One solution is to get the children into school to create a future generation of brilliant engineers who can design machines to manufacture home decorating items at an extremely low cost and with a zero carbon footprint. But this "pie in the sky" dream starts with awareness and definitions of standards. Then, only after decades of continuous improvement, chipping away at the problem month after month, gradually providing more socially responsible options, do we change the world without disruption to social order. That's our dream for this book. We want to start real, near-term action to improve social responsibility performance, rather than idly preach its need.

This shift of focus, from discussion and awareness of the need to be socially responsible to one of action and continuous improvement of social responsibility performance, requires the skill to apply effective tools. One of those tools is risk assessment. Investigating the risk of behaving in a socially irresponsible way, and building systems to mitigate this risk, is our approach to causing the shift of focus. It is through a careful scrutiny of our organizational systems, understanding what it is about those systems that allow members to be irresponsible, and then providing corrective action to the system, that we improve the robustness of those systems toward improved social responsibility. It is a scientific approach to improving human systems.

In this book we are teaching you how to use risk assessment to improve your organization's social responsibility performance. We do not approach social responsibility as altruism, philanthropy, or philosophy. We approach social responsibility as a way for an organization to use scientific methods to improve its performance, and to ensure that the organization is delivering products and services to our children's children's children.

Acknowledgments

We wrote this book in our "spare" time. With full-time jobs, full-time families, and full-time student status, we have been quite busy. Our passion for social responsibility and our experience with business process improvement have compelled us to present this solution. We hear a lot of preaching about the need to be socially responsible. We talk to many people who want to help their organizations take action to be more socially responsible. Out interest and their need is what drove us to carve out time from our busy schedules for this work.

We acknowledge our extremely patient and understanding families for supporting us in this endeavor. Without their help with dinner preparation, watching the kids, mowing the grass, and buying the groceries, we would not have been able to achieve so much. Thank you, to our respective husbands, Gene and Keith, and our children, Jonathan, Isabel, Nazier, and Mekai. Rosemond's parents, Cynthia and Edmond Ragan, were also actively involved in making this book possible.

Author Biographies

Holly Duckworth is a certified Master Black Belt with over 20 years experience in manufacturing. Holly mentors and instructs Black Belts and Green Belts in the Six Sigma tools, training techniques, and influence strategies, with over 2,000 hours as a classroom instructor. She is a member leader for the American Society for Quality and contributes to the Social Responsibility Organization. She has also co-authored *My Six Sigma Workbook*. Holly is currently pursuing a PhD in industrial and organizational psychology at Capella University.

Rosemond (Roz) Moore is a certified Master Black Belt with over 18 years experience in automotive manufacturing. She is an officer for her local section of the American Society for Quality. She has a breadth of experience teaching and leading Six Sigma and Lean Manufacturing projects. Rosemond is currently a faculty member at South Texas College and is seeking a PhD in business at the University of Texas–Pan American. Her research interests include services operations management, human resources, corporate social responsibility, and environmental sustainability.

1 An Introduction to Social Responsibility

What to expect:

- Social responsibility defined
- Social responsibility and the quality movement
- Social responsibility as an ideal
- Social responsibility guidelines (ISO 26000)
- Social responsibility reporting (GRI)

1.1 WHAT IS SOCIAL RESPONSIBILITY?

What is social responsibility? Just walking through the supermarket today, we are bombarded with the "green" label on just about everything. Household cleaning products containing inorganic compounds and bleach are labeled "green." Oil companies advertise their "sustainable" business plans. Some multinational corporations, infamous for their cold-hearted employment tactics, tout social responsibility programs. These terms have lost any merit or validity, if they ever had any. Our readers may ask, if there is no valid definition of social responsibility, isn't it a farce to claim to improve it?

We see social responsibility as an ideal. It is a goal of perfection that will never be completely achieved. It is a target of continuous improvement. It is a business imperative. It is a process-based approach of improvement works best. It is the interconnectedness of organizational actions and decisions toward the impact on the social and ecological sustainability of the community in which it operates.

But these are rather vague, ephemeral concepts. All this talk about what social responsibility is, the awareness and information, does not tell us what to do. We need action; we need tactics. We are embarking on the ability to significantly improve the socially responsible behavior of companies, schools, governments, and other organizations through the definitions and guidelines being built by the International Organization for Standardization: ISO 26000—*Guidance on Social Responsibility.* And if we recognize the similarities of the social responsibility movement to the quality movement, we may be able to accelerate the achievement of our goals.

This book seizes this moment in the history of the social responsibility movement. We take an emerging international guideline on social responsibility, our experience as practitioners of quality improvement tools, and our knowledge of organizational behavior and the management of risk, to develop a tactical tool.

We will be describing the use of a well-worn tool, previously applied by many different professions in the service of mitigating risk—failure mode and effects analysis—and applying it toward the improvement of social responsibility performance. Our hope is to provide a solution to move beyond the discussion of the need for socially responsible behavior, to the ability to just do it. Our goal is to be able to apply the long, difficult lessons learned through decades of trial and error in the quality movement to the emerging social responsibility movement in order to accelerate genuine results toward social and ecological sustainability.

1.1.1 An Important Definition

Before we get too far along in our journey of applying quality tools to social responsibility performance we need to clarify some concepts. We need to make sure we are using valid definitions and guidelines. We may not be able to specify outcomes, we may not want to define compliance specifications, but we can define expectations and intentions. Because social responsibility is an ideal, our definition needs to be a guideline of thought and behavior.

In this book we will be following definitions and guidelines set forth by the Global Reporting Initiative (GRI) and ISO 26000. Later, we'll cover both of these guidelines in detail. There are many other well-recognized definitions. However, these two entities were chosen due to their global acceptance and, in the case of ISO 26000, the global consensus used during development. Significant struggle and deep contemplation were required to achieve these global guidelines. The inclusion of a diverse, global group of thought leaders in the development of these guidelines has warranted our respectful use of them.

Let's begin with the definition of *social responsibility*. Our favorite is excerpted from ISO 26000: "the responsibility of an organization for the impacts of its decisions and activities on society and the environment, through transparent and ethical behavior" (Figure 1.1). Upon close investigation, we can recognize that this is an incredibly complex definition. For example, it states a responsibility, not a

---

The responsibility of an organization for the impacts of its decisions and activities on society and the environment, through transparent and ethical behavior that contributes to sustainable development, health and the welfare of society; which takes into account the expectations of stakeholders, is in compliance with applicable law and consistent with international norms of behavior, and is integrated throughout the organization and practiced in its relationships. This includes products, services and processes. Relationships refer to an organization's activities within its sphere of influence.

This definition includes the following critical concepts:
- A voluntary acceptance of accountability
- A responsibility of actions and the impact of the outcomes of those actions
- An ideal sought first through careful decision making
- The interconnection of society and the environment
- The importance to behave ethically and transparently.

FIGURE 1.1   A definition of *social responsibility* (from DRAFT ISO 26000 WD4.2).

requirement. This speaks to social responsibility as an ideal, a voluntary acceptance of accountability. Next, this definition requires our attention not just to actions and decisions, but to the impacts of those actions and decisions. We must be wary of what our organization does, as well as the outcome of the actions. The proper outcome must be achieved. An intention, an action, and an outcome are encompassed within the meaning of social responsibility. Next, the term *decision* is included in this ISO 26000 definition of social responsibility. At the root of striving for an ideal is a way of decision making. How we make decisions will become important in our social responsibility journey. The next important words in the ISO definition are *society* and *environment*. These entities are interconnected. Society cannot exist without the environment. The definition, by including both words, makes this interaction explicit. As we will see in later chapters, constructs such as human rights, ethical governance, and social development are as important as ecological protection in the full context of social responsibility. It isn't only about "green," but it's also about the maintenance of a social fabric that can protect the environment. And lastly, we like this definition because it tells us not just what to do, but how to do it—through transparent and ethical behavior. Achieving beneficial social and ecological actions and decisions without transparency and ethics is not social responsibility.

As an ideal, the best social responsibility performance is constantly changing. Seventy years ago, we were not aware of the hazards in lead paint. Thirty years ago we were not aware of the hazards of dioxins used in polyvinylchloride plastic production. Today, we have not yet proven if the chemicals used in plastic manufacturing are related to a rise in autism spectrum disorders. Acceptable behaviors constantly change. Knowledge is gained. We may be ignorant of our own ignorance. And because of our recognition of our impact and accountability to society, transparency and ethical behaviors are needed. The principles that govern appropriate conduct and concern for others become an integral part of the tactical deployment of an improvement initiative that attempts to achieve an ideal.

We will carry this definition forward for the remainder of the book. We will utilize the emerging guideline from the GRI and ISO to tailor our work, and we will draw from our experience utilizing effective product and process quality improvement tools. We will take our experience and these guidelines and demonstrate tools for you to use to achieve outstanding performance in social responsibility for your organization.

1.1.2 SIMILARITIES TO THE QUALITY MOVEMENT

The quality movement is a reference to a period in the mid 20th century when business management philosophies began to employ systems of product and process quality assurance. This period began with Walter A. Shewhart's deployment of statistical quality control techniques in the 1920s, moving to W. Edwards Deming's involvement in product quality improvements during World War II and in the postwar Japanese rebuilding, and then to A. V. Feigenbaum's Total Quality Management concepts adopted in the 1980s. The quality movement is marked by the continued embedding of scientific methods of process and product quality control and organizational culture adaptations of quality management systems. The most recent elements of the quality movement are the utilization of Lean Production and Six Sigma continuous improvement efforts.

We liken the current state of the social responsibility movement to the early days of W. Edwards Deming and the quality movement. Just as Deming, in the 1950s and 1960s, was changing the attention of industry to focus on product quality and process control (Figure 1.2), the current trend toward formalizing methods to ensure socially responsible business management is changing the focus of attention today. Deming began his path by advising businesses that had made large fortunes by selling poor-quality products. The consumer had no choice. It was the time of *caveat emptor*: Let the buyer beware. That is not to say there aren't still worthless products being hawked, but consumers have a very different expectation of purchased quality now than they did in the 1940s and 1950s.

W. Edwards Deming, Joseph Juran, Taichi Ohno, and other quality movement activists achieved this culture change by working with just a few companies who would listen. Those companies used certain analytic and management methods to deliver breakthrough improvements in product quality. Many, such as Toyota Motor Corporation, have become famous and are now mimicked for their organizational culture, which is devoted to product quality and process improvement. The products of these companies took market share, significant market share in the case of Toyota, from their competitors. If the competitors wish to stay in the market, they too must direct their focus on product quality. Gradually, over decades, a foundational market shift has occurred. Quality assurance is now a fundamental aspect of business management. High product and service quality is a business imperative.

It is our contention that the same market forces of change are just beginning with social responsibility. Very few businesses have a focus on operating in a socially

---

1. **Appreciation of a system**
   This includes the recognition that all factors of influence have the potential to interact with each other. People interact with processes, products interact with the delivery of service, processes interact with the environment, etc.

2. **Knowledge of variation**
   This includes an understanding of how things "go wrong." The robustness of a system against defects and how close to the boundary failure the process is operating are included in the knowledge of variation.

3. **Theory of knowledge**
   This includes the idea of not acting before studying. We cannot know the best course of action unless we understand how to study the system. The plan-do-study-act method of problem solving is an example of the theory of knowledge.

4. **Knowledge of psychology**
   This is the recognition that all systems of humankind require the knowledge of how people interact with the system. Knowing the mechanics of a system is not enough; one must also understand the motivations and cognitions of the people interacting with the mechanical system.

FIGURE 1.2   W. Edwards Deming's system of profound knowledge.

responsible manner. But, as before Deming, there is no current standard, no tests of validity, no direction for improvement, and no minimum expectations. All we have are an impending guideline, a few pervasive definitions, a few awards for superior performance for social responsibility, and some advertising hype. Today, in the quality movement, we have formal programs such as Total Quality Management, Six Sigma, Capability Maturity Model Integration, and Good Manufacturing Practices. We have ISO 9000, AS 9100, and TS 16949, just for a short list, as formal standards and guidelines for quality management systems. Shewart, Ishikawa, Juran, Feigenbaum, Taguchi, Crosby, Ohno, Shingo, and many more are thoroughly studied masters of sage advice in the field of quality assurance. We have exemplars in product and service quality such as Toyota, Nordstrom, Deere, Proctor & Gamble, and BMW. We have extended warranties and product liability laws. And in the United States, for example, we have the Malcolm Baldrige National Quality Award. An entire profession and professional industry has arisen, moving the marketplace from an attitude of "let the buyer beware" to the realization that product and service quality is a business imperative to be leveraged for market share gain.

It is our prediction that in a few short years, the same will be said of social responsibility. Standards and guidelines such as the Global Reporting Initiative and ISO 26000 are being deployed. Some corporate exemplars such as IKEA, UPS, International Paper, and Starbucks are gaining recognition for their social responsibility actions. We have a growing list of recognition and award programs (Figure 1.3). Gradually, a profession and professional industry is emerging. However, there are still more questions than answers surrounding *how* to improve, not just talk about, social responsibility. There are many companies who genuinely wish to improve their social responsibility performance, but simply don't know how to take action. That is our mission with this book.

We wish to move beyond talking, advertising, and creating public relations press releases on social responsibility performance. Our goal is to provide tools, well-worn tools, shamelessly borrowed from the quality movement, to apply to the genuine improvement of social responsibility. There are scientifically proven, valid instruments, easily applied toward process improvement, which with only a slight change of focus, can be applied to an organization's social responsibility initiative. This book is focused on only one of these tools: failure mode effects and analysis (FMEA). As the social responsibility movement matures, we predict more and more of these well-worn quality management tools will be effectively applied to social responsibility performance improvement.

Before we leave the comparisons to the quality movement, there is another critically important similarity between quality and social responsibility. That is the idea of an ideal. There are many other initiatives in which companies engage. For example, in the United States, financial controllers must lead Sarbanes–Oxley compliance

- Green Power Partnership—EPA
- World's Most Ethical Companies—Ethisphere
- Corporate Social Responsibility Index
- Fortune's Most Admired Companies
- Fortune Global 500

FIGURE 1.3 Social responsibility recognition and award programs.

assurance initiatives. Many of our information technology departments are engaged in Capability Maturity Model Integration deployments and TL 9000 certification. And of course, many employee safety initiatives are well managed and doing great things. However, both quality and social responsibility differ from these types of performance improvement programs. Most of these other programs are initiated, deployed, and measured based upon the goal of compliance. Are the laws being followed? Are the project tasks being accomplished? Are the correct behaviors being observed? The outcome of these programs can be measured as either in or out of compliance. We are either abiding by the legal requirement or not. We are either certified or not. We are either hurting employees or not.

For both quality and social responsibility, the end goal is an ideal. We begin the journey in full recognition that perfect quality will never be achieved. The customer is always demanding improved quality, and competitors are always achieving it. For social responsibility, we know that we will always have some negative environmental and social impact. And we know that we will never reach a resting point or end goal. The minute one level of performance is achieved, continual improvement to the next level is immediately expected. The stakeholders are always demanding improved responsibility performance, and competitors are achieving it.

Quality is an ideal. Social responsibility is an ideal. It is this attitude that compliance is not nearly enough that makes social responsibility initiatives the perfect candidate for application of many quality improvement tools. And because these concepts are ideals, the definitions are always shifting. What was perceived of as high quality three decades ago is poor quality today. What might be great quality for one customer may be unsatisfactory for the next customer. For example, the expectations of the recycled content of product packaging is rapidly changing at this moment, and actions that may be considered the norm of social responsibility in Finland may be highly irresponsible in Zimbabwe. The achievement of an ideal is like grasping at smoke. You can feel it. You know it when you see it. You know when it's there. You know when it's not there. But it's really tough to define it in hard, specified outcomes—and impossible to define as merely compliance to policy.

1.2   AN INTRODUCTION TO ISO 26000

ISO 26000, *Guidance on Social Responsibility*, is an effort begun through actions initiated as early as 2001 by the International Organization for Standardization (ISO). This guideline is intended to be a globally consistent, practical guideline for any organization wanting to enhance its social responsibility performance. It is the basis for our approach to this book. The ISO has targeted publication of ISO 26000 for late 2010. There are about 90 countries and 40 organizations that are participating in the development and future publication of this social responsibility standard. The standard will address what elements for us are important for an organization to operate in a socially responsible manner. It is important to note that it is being published as a guideline, not a standard of certification requirements. There is no expectation that third-party certification to ISO 26000 will take place. As a guideline it is intended to do just that—provide guidance.

The content of ISO 26000 is focused on internationally standard definitions and terminology associated with social responsibility. The ISO 26000 document provides guidance on key social responsibility elements and best practices. The pressure placed on businesses to operate in a socially responsible manner is addressed in ISO 26000 from a multiple stakeholder perspective. These six stakeholder groups (industry, government, labor, consumers, nongovernmental organizations (NGOs), service/support/research/others) are represented as members of the drafting task force, which is responsible for building the ISO 26000 drafts.

The effort, argument, discussion, and consensus required to bring the entire world together to develop a single guideline on social responsibility is daunting. By the time of its formal publication, these diligent contributors will have invested over five years toward its creation. The difficulty of writing a guideline on, for example, labor relations, as applicable in China as in the United Kingdom, is nearly overwhelming. But it is critical that a foundational document be developed. We cannot know if we've improved social responsibility performance if we have no basis upon which to measure the performance. Having an international guideline is a critical first step in an effort to genuinely improve social responsibility performance.

There are seven key social responsibility subjects that were agreed upon during the ISO 26000 working group meeting held in Sydney, Australia, in 2006. These seven core social responsibility subjects are organizational governance, human rights, labor practices, the environment, fair operating practices, consumer issues, and community involvement and development (see Figure 1.4). Further development and explanation for each of the core subjects will be provided in the following section. In addition to the core subjects of social responsibility, the ISO 26000 working group has also specified seven key principles of behavior: accountability, transparency, ethical behavior, respect for stakeholder interests, respect for the rule of law, respect for international norms of behavior, and respect for human rights (Figure 1.5).

- Organizational governance
- Human rights
- Labor practices
- The environment
- Fair operating practices
- Consumer issues
- Community involvement and development

FIGURE 1.4   The seven core subjects of ISO 26000.

- Accountability
- Transparency
- Ethical behavior
- Respect for stakeholder interests
- Respect for the rule of law
- Respect for international norms of behavior
- Respect for human rights

FIGURE 1.5   The seven key principles of ISO 26000.

The core subjects provide the detail of discrete elements of an organization's responsibility. There is some overlap between the seven core subjects; for example, some human rights issues surely overlap with labor practices. However, the guideline has been written in a manner to provide detailed information of expected decision making and activities of focus for each core subject. The seven key principles of behavior are intended to span all of the core subjects. These are guidelines of behavior expected in any, and all, subject areas.

Think of the seven core subjects as the issues upon which decisions, impacts, and actions are to be taken and the seven principles of behavior as how those decisions and actions should happen. A holistic social responsibility improvement initiative will consider both the subjects and the principles. We can't leave any of the seven core subjects or seven principles behind. And we can't just focus on the *what*; we need to simultaneously focus on the *how*.

Publishing an international guideline on social responsibility is critical, but a very small first step. We recognize the need for guidelines of socially responsible behavior, but this is no guarantee that organizations will behave in this fashion. This is evidenced by the negative impacts of irresponsible behavior and has accelerated concerns associated with the lack of social responsibility. The call to action needs to begin with a solid foundation of understanding the current state. In order to know if our process changes are in the direction of improvement, rather than degradation, we must also know what the improved state looks like. We need standards and guidelines to tell us how bad is bad, how good is good, what to consider, and of what risks to be aware. We believe this is achieved through an internationally recognized guideline, such as ISO 26000.

1.3 THE GLOBAL REPORTING INITIATIVE

Beyond ISO 26000, another important international guideline on social responsibility is the Global Reporting Initiative (GRI). The GRI provides guidelines that have been accepted worldwide for the reporting of sustainability results. The intent of the commission that created the GRI was to provide a framework that encouraged organizations to routinely report on sustainability results just as they do with financial results. The goal of sustainability as defined by the World Commission on Environment and Development (WCED) is "to meet the needs of the present without compromising the ability of the future generations to meet their own needs." The original framework for the GRI was developed in 1997 through the collaboration of key stakeholders such as nongovernmental organizations, businesses, investors, labor unions, and others in a global context and using a consensus approach. This collaboration led to the creation of the GRI reporting framework, which defines what to report (standard disclosures and sector supplements) and how to report it (principles, guidance, and protocols).

The GRI is currently in its third revision, called G3, which was launched in October 2006. The intent is to continuously update and improve on the framework to serve the ever-changing business environment. The GRI is considered the most utilized global standard for sustainability reporting worldwide due to its support from the United Nations Environment Program (UNEP) and its cooperation with the United Nations Global Compact. It is considered credible because of the global, multistakeholder development of the framework. The GRI is governed by a board of directors, stakeholder council, technical advisory committee, organizational stakeholders, and a secretariat. The GRI is the result of work conducted by a not-for-profit organization based in Amsterdam. It has since grown to a public report of over 900 participating companies, which use the GRI sustainability reporting framework to bring their corporate reporting into compliance with these expectations.

This framework of reporting is intended to provide consistent guidance on the reporting of sustainable economic, environmental, and social performance equal in validity to a company's reporting of its financial performance. Today, it is a rare corporate annual report that doesn't have some paragraph on the company's vision and performance of social responsibility. However, without the GRI framework, these reports have no standards on what will be reported and how it will be reported. The GRI begins to develop scaffolding upon which companies can transparently display their social responsibility challenges and accomplishments. The GRI provides guidelines, protocols, and industry sector supplements. And it provides an extensive library of report guidelines.

The GRI is not yet as well known in the United States as it is in Europe. At this writing, only 100 of the 900 reporting GRI organizations are headquartered in the United States. The GRI categories of reporting are economics, environment, human rights, labor practices, product responsibility, and society (see Figure 1.6). The G3 guidelines go into great detail on how to define, measure, calculate, and report on over 60 performance indicators within these categories. Each reporting organization lists a self-report in an Internet database, which can be viewed by other reporting organizations. An organization can also have the GRI check their report for compliance with the reporting guidelines. This provides transparency of the social responsibility actions across all performance indicators to stakeholders. Companies can also begin to benchmark others through the reporting.

In summary, we find that ISO 26000 is best used as a guideline for defining decision making, actions, and behaviors of social responsibility. The GRI is best used as

---

- Economic: 9 indicators
    Example: Range of ratios of standard entry-level wage compared to local minimum wage at significant locations of operation.
- Environmental: 30 indicators
    Example: Percentage of materials used that are recycled input materials.
- Social/labor practices: 14 indicators
    Example: Percentage of employees covered by collective bargaining agreements.
- Social/human rights: 9 indicators
    Example: Percentage and total number of significant investment agreements that include human rights clauses or that have undergone human rights screening.
- Social/society: 8 indicators
    Example: Total number of legal actions for anticompetitive behavior, antitrust, and monopoly practices and their outcomes.
- Social/product responsibility: 9 indicators
    Example: Total number of substantiated complaints regarding breaches of customer privacy and losses of customer data.

FIGURE 1.6 The Global Reporting Initiative summary of performance indicators and examples.

a way to measure and transparently report on the outcomes of those improvements. From here we will be focusing on the ISO 26000 elements, which are slightly different from the GRI categories. We do not mean to diminish the power, or our support, of the important work accomplished through the GRI. We will use ISO 26000 to guide our risk mitigation work, but we would also encourage any organization to participate in the GRI to provide valid transparency to its stakeholders.

1.4   THE SEVEN CORE PRINCIPLES OF ISO 26000

Our most esteemed respect is given to the participants in the creation of the ISO 26000 guideline. The importance of having a globally consistent standard of social responsibility is critical. As we can imagine, each of the representative countries and individuals brings a unique focal area and set of issues. Coming to consensus on standard guidelines of human rights between members from Bangladesh and Finland can only be a labor of love! It is also our hope that by writing our book, and designing our tool for failure mode effects and analysis around the core issues detailed in ISO 26000, that we help to promote this important work.

The detail provided in the ISO 26000 guideline is extensive. There are seven principles of behavior that span all seven core subjects. And then, within the seven core subjects there are details on primary issues. These issues are intended to be thought provokers. They are detailed because they are the most probable social responsibility dilemmas in organizations today. The issues listed (see Figures 1.7 to 1.13) are not intended to be exhaustive; they are not intended to allow the organization to ignore issues that may be unique to that business. They are intended to be the starting point of investigation toward issues for which stakeholder impact needs to be considered. In the next section, we will go into further detail on the seven core subjects and their associated issues.

1.4.1   SOCIAL RESPONSIBILITY PRINCIPLES AND EXAMPLES

The call to action needs to begin with a solid foundation of understanding the current state. In order to know if our process changes are in the direction of improvement, rather than degradation, we must also know what the improved state looks like. We need standards and guidelines to tell us how bad is bad, how good is good, what to consider, and of what risks to be aware.

The term *stakeholder* is an important concept with respect to social responsibility. Stakeholder is defined, in ISO 26000, as an individual or group that has an interest in any activity or decision of an organization. This is much broader coverage of interest than customer or shareholder. One of the first steps of behaving in a socially responsible manner is for an organization to identify and recognize its body of stakeholders. For example, in a typical manufacturing organization, easily recognizable stakeholders would include employees, customers, and stock shareholders or company owners. However, the local community, the local, regional, and national government, charitable organizations, professional societies, institutions of education, suppliers, energy providers, fire protection and police agencies, and spouses and children of employees, among many others, comprise the body of stakeholders. Each of these groups has an interest in the activities and decisions of the organization. Some of the interest is due to the ability to be harmed by irresponsible actions and decisions; some of the interest is due to the ability to benefit by the responsible actions or decisions. Acting with social responsibility starts with the recognition of this influence on the larger network of stakeholders.

There are seven key social responsibility core subjects. The seven core subjects (organizational governance, human rights, labor practices, the environment, fair operating practices, consumer issues, community involvement and development) are also supported by specific issues that are addressed under each. In addition to the core issues of social responsibility, the ISO 26000 working group has also specified seven key principles of social responsibility. These key principles (accountability, transparency, ethical behavior, respect for stakeholder interests, respect for the rule of law, respect for international norms of behavior, and respect for human rights) are part of the supporting structure of the social responsibility guidelines (Figure 1.5).

These seven principles span across all six stakeholder groups and all seven core subjects in importance. Every socially responsible action or decision will be made in accordance with these seven principles. When defining what socially responsible behavior is, the first step is to recognize the broad fabric of stakeholders, and the second step is to conduct actions and make decisions in consideration of these seven principles. In the following sections we will detail the meaning of these principles, demonstrate public examples of these principles, and explain how their importance could be utilized in an initiative to improve social responsibility performance.

1.4.1.1 Accountability

According to the ISO 26000 guideline, accountability refers to an organization's response to social and environmental issues. The guideline suggests that the organization should be held accountable not only for its decisions and actions related to social and environmental issues, but also for the impact of those issues on society as a whole. Examples of companies that are socially responsible can be found in the Fortune 100 Accountability Ranking. Accountability is a three-step response to negative outcomes. When negative outcomes arise, an accountable organization will take responsibility for the outcome, remedy the outcome, and instill preventive measures surrounding the potential recurrence of the outcome. Accountability is recognition that no organization is perfect, but we must trust that appropriate action is taken when problems arise.

1.4.1.1.1 Example of Corporate Accountability

Starbucks is often mentioned when conversations of social responsibility are undertaken. Starbucks from the very beginning has strategically structured its operations to be socially responsible. Starbucks takes steps to ensure its accountability by requiring suppliers to extensively document their environmental and social actions taken to produce coffee. There is trust in appropriate action. Starbucks also demonstrates a level of accountability through the offering of microcredit loans to farmers in Latin

America to cover preharvest costs. Starbucks has received much recognition for its social responsibility efforts. It has been identified by Fortune as one of the "100 Best Places to Work" (2000–2008). It has also been recognized as one of the most admired companies (Fortune, 2003–2007), a "100 Best Corporate Citizens" (2000–2007) from two separate entities (*Corporate Responsibility Officer* magazine and *Business Ethics* magazine), and one of the "World's Most Ethical Companies" (Ethisphere, 2007), to name just a few accolades. While Starbucks is considered a very accountable firm, it has been criticized in some circles for the need to improve on its transparency with regards to the reporting of the amount of pesticides used by its suppliers, an example of an organization excelling at one principle but lacking focus in another.

One very familiar example of accountability is the 1982 Tylenol-tampering scare, which has become a case study of how to handle a crisis. In this example, bottles of the over-the-counter pain medication Tylenol were tainted with cyanide after they had been placed on the store shelf. Although Johnson & Johnson, the makers of Tylenol, were not at fault for the tampering, they realized the weaknesses of their product packaging to prevent this horrible outcome. They took remedial, proactive, and preventive measures by changing to tamper-resistant packaging. This is an example of an organization demonstrating accountability to stakeholders even when the source of a hazard is outside of the organization's cause.

As we will discuss in later chapters, when an organization embarks on an initiative to improve its social responsibility performance, the whole value stream must be analyzed for potential risk. And the full body of stakeholders should be recognized while identifying these risks. The risk of potential failures of products and services to any stakeholder should be reduced. Taking responsibility for the risks inherent in products and services, acting to reduce these risks, and taking preventive action is the basis for accountability and improving social responsibility by mitigating risks through failure analysis.

1.4.1.2 Transparency

An emphasis on transparency demonstrates how well the organization communicates or makes information available about its practices and the practices of its key partners and stakeholders. It is the act of conducting business actions and making decisions in a nonsecret environment. Transparency involves open access by stakeholders to the information and process involved in organizational decision making. Transparency is achieved through the accessibility of the company to its stakeholders in times of crisis or need. Companies can demonstrate their level of transparency by providing toll-free numbers to allow consumers access to information about the company at all times. A transparent organization also has a candid disclosure system ready for execution in the case of a customer issue. Transparency also takes the form of reporting proactively without legal impetus and reporting as required by government regulations.

After a string of public debacles, such as financial issues (Enron, WorldCom, etc.), sexual misconduct (the Catholic Church), world health concerns (SARS), and war reporting (the first Iraq war), the general public began to demand organizations, governments, and other entities to be more forthcoming and transparent with regard to their actions and decisions. Some progress has been made in this area, such as in the United States with the Sarbanes–Oxley Act, which requires full disclosure of off-balance-sheet activities and certification of financial reports. The Global Reporting Initiative (GRI) is also a promising movement toward transparency. When considering examples of good transparency, two companies can be often found at the top of the list, Ben & Jerry's and The Body Shop.

1.4.1.2.1 *Example of Corporate Transparency*

From the very beginning Ben & Jerry's emphasized measuring performance not only financially, but also socially and environmentally. Ben & Jerry's published their first social performance report in 1989 and were openly candid about their inability to make change in the communities where they operated. The Body Shop also openly communicated its desire as an organization to be socially responsible. This emphasis of openly communicating the actions and decisions of the organization and their impact on social and environmental matters was a novelty at the time. Both were seen as pioneers in their industries and, while supported as being open, were also able to be challenged by competitors who also had knowledge of their positive successes and failures at social responsibility. Their groundbreaking approaches, leading to market success, may have been the root of organizational demise in both of these case studies. Ben & Jerry's was purchased by Unilever, which some argue has not been able to maintain the values of Ben & Jerry's. The Body Shop was eventually purchased in 2006 by L'Oreal, another large corporation, as economic changes drove smaller firms out of business.

Our recommendation to mitigate the risk of social responsibility failures through risk analysis creates documents upon which risks are identified, and requires actions and decisions on which of those risks will be mitigated. Not all risks will be mitigated. There will be a priority within the risk analysis. This will require organizations embarking on a social responsibility performance improvement initiative to prepare for the transparent dialogue with its stakeholders on these decisions and actions. Some stakeholders may not be immediately pleased. But the principle of transparency requires that the organization be prepared to share the information surrounding decisions, including decisions of risk mitigation, and to communicate the intentions and information resident in the decisions and actions.

1.4.1.3 **Ethical Behavior**

Ethical behavior involves the organization's internal practices and procedures and how standards or codes of conduct are established. Ethical behavior includes acting with integrity, honesty, fairness, and concern for all stakeholders and the environment. Ethical behavior is a commitment of acting in the best interest of all stakeholders. Most are aware of the multitude of examples of unethical behavior, as exhibited by Enron, WorldCom, Tyco, and more recently, Bernie Madoff. Each of these examples demonstrates cases where leadership within an organization participated in, overlooked, and encouraged practices within the firm that were unethical. These were actions and decisions that benefited some stakeholders at the expense and harm of others.

1.4.1.3.1 *Example of Ethical Behavior*

A ranking of the most ethical companies is conducted by Ethisphere Institute. Its 2009 ranking has identified such companies as General Mills, Caterpillar,

Honeywell International, Nike, BMW, and Toyota Motors. Its ranking of the most ethical companies includes various elements of social responsibility weighted in the following manner: corporate citizenship and responsibility (20%), corporate governance (10%), innovation that contributes to public well-being (15%), industry leadership (5%), executive leadership and tone from the top (15%), legal, regulatory, and reputation (20%), and internal systems and ethics/compliance programs (15%).

General Mills' Executive Vice President, General Counsel, Chief Compliance and Risk Management Officer speaks about how the company uses telephone calls to the ethic's hotline as lessons learned, which are shared with all employees. He also discusses the importance of having an ethical and compliance program that engenders the values of the company's culture. Caterpillar's chief ethics and compliance officer (CEO) discusses how the values, beliefs, and actions must come from the top (from the Chief Executive Officer). Many aspects of ethical behavior involve ensuring that the actions of the organization are not harmful to its stakeholders. An ethics program is only as good as the leader who supports it and promotes it.

Embarking on a social responsibility performance improvement initiative will result in the change of behaviors. An organizational climate that encourages the open discussion of ethical dilemma and the reporting of potential unethical behavior will improve adherence to this principle. It is important for leaders to model ethical behavior. Opportunities to demonstrate and communicate incidents of concern for all stakeholders are important for leaders. If all organizational constituents see their leaders communicating their own decision making with respect to integrity, honest, and fairness, they will be more likely to take these same principles into consideration with their own decisions.

1.4.1.4   Respect for Stakeholder Interests

Just in case you haven't yet understood that stakeholders are important to social responsibility performance, ISO 26000 has designated a principle dedicated to the respect of stakeholder interests. Since each of he stakeholders plays a key role in an organization's social responsibility, it is important to ensure that all stakeholders are identified and their interests are being met. Consideration for the various stakeholders of a firm opens the firm up to thinking about how its organizational actions impact not only internal stakeholders (suppliers, employees, stakeholders, etc.), but also external stakeholders (consumers, government, NGOs, community, etc.). This may be much easier said than done. Some stakeholder interests may conflict with others. If the firm's key focus is on financial performance, many times the decisions made can negatively impact external stakeholders. Or a decision by an organization to engage in collaboration with an international professional society may preclude the same resources being dedicated to local community activities.

1.4.1.4.1   Example of Stakeholder Interest Behavior

Wal-Mart, while in the news recently with regards to socially irresponsible behavior (unethical pricing practices that shut out small-town retailers, denying employees payment for overtime work, underpayment of employees, etc.), has made a personal commitment to social responsibility and is ranked by Fortune as a "Top 100 Accountable Companies" (2008). Since it only manufactures 8% of what it sells, Wal-Mart has taken on the task of convincing its suppliers to be more socially responsible, according to David Blackwell, Wal-Mart vice president and chief financial officer (CFO) of global procurement. This emphasis on involving the supplier in socially responsible practices can have a major impact on society when a company the size of Wal-Mart is actively involved. Wal-Mart was able to change public opinion on its social responsibility stance when Hurricane Katrina hit New Orleans, Louisiana, in 2005. It was able to save lives by rapidly using its strong supply and employee network to deliver food, medicines, and materials for shelter when the government was unable to do so.

If members of an organization are behaving with accountability, transparency, and ethics, they are already respecting some of the interests of their stakeholders. However, this particular principle goes beyond just accountability, transparency, and ethical behavior. This principle is a recognition that without stakeholders, and their interests, the organization cannot exist. Without customers there is nothing to sell. Without constituents there is nothing to govern. Without individual members there is not an organization. The stakeholders and the organization are inextricably linked. Respecting this dependency of interests is a key principle of social responsibility.

1.4.1.5 Respect for the Rule of Law

The laws of a city, state, and country where an organization operates must be taken into consideration and followed as a practice. This requires the organization to be aware of the laws in the location in which it operates, and to audit its compliance to those laws. This principle carries the intention of the organization to also follow its own policies and procedures. It is a principle of respect for those procedures that have been designated as appropriate by the ruling organization. Now, this may also be one of those guiding principles that it is easier said than done. Unfortunately, there are many countries in which apparently unethical laws still exist, such as those allowing the ownership of women. Or the intentional absence of law may allow for unethical behavior, such as those countries that do not prevent child labor. Or the common law that expects socially irresponsible action, such as those who sponsor bribery and corruption. However, this principle of social responsibility requires the recognition of formal channels upon which these irresponsible behaviors should be addressed. It is not acceptable to intend to behave in a socially responsible manner by breaking the law. It is acceptable to work toward changing irresponsible laws.

1.4.1.5.1 Example of Rule of Law Behavior

The respect for the rule of law becomes extremely important as companies expand globally. This expansion leads organizations to consider if business practices should be standardized to those that exist at the headquarters or should be implemented to fit the law of the city, state, or country in which the operations exist. The rankings by Ethisphere include a 20% weighting for the company's legal, regulatory, and reputation record. This ranking demonstrates the importance of companies conducting themselves to follow the rule of law. While many instances of the illegal act of insider stock trading can be found with ready identification of the perpetrators, we must rely on rankings and the lack of accusatory newspaper headlines to identify those firms with legally sound practices.

One example of this principle is AstraZeneca's global policy on legal and intellectual property. Its transparent policy, as found on its public Web site, states that all company staff (employees and contractors) must take steps in all of their activities to comply with the national and international laws, regulations, and codes. The statement includes requirements that staff must follow the law, not condone others to disobey the law, be aware of the laws or know where to find them, and involve the legal department in any matter as appropriate. Other companies that take a similar approach are Coca-Cola and Nestlé Corporation.

When utilizing an approach of risk mitigation through failure analysis for social responsibility performance improvement, the risk of breaking the law should be identified. Respect for the rule of law should include those laws with which the organization is led, both internal and external, as well as those laws in the local communities in which the organization operates. Recognition of regulatory agencies as stakeholders is the beginning of adherence to this principle. Care should be taken when the organization finds its internal principles of ethical behavior in conflict with the local law. The organization should use its influence to affect statutory change and not be tempted to take compliance into its own hands. In the long run, an organization that abides by its own set of laws and does not work to make the laws better for all constituents is operating in a socially irresponsible manner.

1.4.1.6   Respect for International Norms of Behavior

International norms of behavior must be followed as organizations become more global. It is not good enough to only follow the norms of the headquartered country, but also all other countries in which a company operates. Some might see this as a contradiction with the law. For example, in Mexico it is acceptable to provide financial incentive to take on a business deal, but in the United States this is illegal. In other words, just because a local norm allows for bribery, in a legal sense, it is not a respected international norm of behavior. Most firms make it very clear that no activity considered illegal internationally should be overridden in the case of international norms of behavior. However, the choice for a state or country to follow international norms may be the result of coercion, cooperation, coordination, or coincidence of interest.

1.4.1.6.1   Example of International Norms of Behavior

The current financial crisis in the United States is an example where, in many instances, no laws were broken, but the country norm of giving home loans to individuals who could not afford the payments has had a major international financial impact. Whether or not to follow the international norms of behavior is left up to the organizational leaders' decision making. These leaders must not only decide whether a behavior is legal, but also keep in mind that international behavior must be given consideration. ISO 26000 gives guidance that the international norm should be favored over the local practice where the international norm sponsors sustainable development and improved social welfare. Clearly the subprime mortgage debacle of 2008 has not promoted sustainable development.

The most egregious examples of a lack of respect for the international norms of behavior are those of rogue states, such as North Korea. These rogue entities are marked by self-isolation, obsessions, and tyranny. Although the nation of North Korea is easily recognized as not abiding by the international norms of behavior, it is important to note that other entities, such as businesses and organizations, can also be guilty of ignoring the international norms of behavior. And because international norms of behavior do not go hand in hand with legal behavior, one must be cautious. An example of this crossover between state and international behavior can be found relative to legal proceedings associated with the National Free Trade Agreement (NAFTA). NAFTA established the free movement (nontaxed) of materials between Canada, the United States, and Mexico for manufacturing. What right does UPS have to deliver packages directly to locations in Canada when Canada has a governmental provision that only the Canadian postal service can provide deliveries? Can UPS challenge the norm of behavior for Canada given NAFTA? In the same token, can a Canadian company challenge a "Buy American" law for highway construction projects in the United States?

The conflict between locally acceptable actions and decisions and those of respected international norms is at the heart of the social change involved in an organization's social responsibility performance improvement initiative. If the organization operates in localities in which social progress has not yet met international norms, e.g. third-world countries, the organization should prepare for some very tense actions and difficult decisions. It is not without strife that local culture changes. And what may appear as a long-term benefit for the society as a whole may be a crisis-initiating detriment to a few in the current culture.

1.4.1.7   Respect for Human Rights

The internationally respected rights of all humans, where utilized for the purposes of business, must be followed. The definition of human rights is recognized through the United Nations International Bill of Human Rights. It includes the admonition of discrimination, torture, kidnapping, slavery, the abuse of children (in armed conflict, the sale of, and the use in sex trading), and the abuse of migrant workers and those with disabilities. Ensuring that persons involved in the execution of business activities are treated with respect to their full human rights is considered to be universal. There are many news claims with regards to toy and clothing manufacturers using labor that is in violation of human rights. A company's responsibility for protecting human rights exists in all countries in which it operates. While ISO 26000 provides the guidelines for appropriate human rights behavior, the GRI in 2008 made a call to action for firms to become involved in actively participating in defining and reporting on human rights activities.

1.4.1.7.1   Example of Human Rights Behavior

GE committed in 2007 to take a more active role in human rights in its operations in 2008. Key actions undertaken by GE were the implementation of human rights guidelines, the inclusion of human rights principles in its supply chain, and influencing direct business partners to be more actively involved in human rights issues. An established organizational position (vice president of corporate citizenship) within the organization has the task to assess and ensure that all business entities within GE are reporting human rights infractions. In this role, this vice president is responsible for analyzing human rights concerns and ensuring compliance to the established guidelines. With regard to its supply base, GE conducted a benchmarking exercise to determine what aspects of the International Labor Organization (ILO) human rights declaration should be emphasized. It determined through reviews with other companies that its main focus at this time would be increasing supplier awareness. The activities initiated with the direct business partners with regards to human rights involve the sharing of worldwide protocols used for advancing human rights, including assessments or audits to ensure that human rights abuses are not conducted. GE is a perfect example of a company that has shown commitment not only through words, but also through its actions.

Most companies of such a progressive nature as needed to initiate a program of social responsibility performance improvement are well beyond the careful concern of human rights for their employees and customers. However, some progressive companies have found egregious actions far upstream in the supply chain. Are the diamonds found on your machine tools a result of slave miners? Is the raw wool in your garments sheared by children? When assessing risk of violations of human rights, it will be necessary for organizations to evaluate their whole supply chain, all the way back to raw materials.

1.5   THE SEVEN CORE SUBJECTS OF ISO 26000

Social Responsibility is defined in ISO 26000 as "the responsibility of an organization for the impacts of its decisions and activities on society and the environment, through transparent and ethical behavior." Contingent with this definition are seven core subjects defined by ISO 26000; advice is given by the ISO guideline for an organization to examine its issues in these core subject areas in order to improve social responsibility performance. These seven subjects are not intended to be exhaustive or exclusive, and for some organizations one or more of these core subjects may not apply. They are intended as a foundation of globally consistent areas of focus for improvement, and thus are ideally suited as a starting place for us when building a tool to mitigate risk and continuously improve social responsibility. We will take each of these core subject areas and briefly describe the issues that should be examined as your organization begins to work on improving social responsibility.

1.5.1   ORGANIZATIONAL GOVERNANCE

Organizational governance speaks to how an organization makes decisions, including the decisions associated with how the organization is structured. The primary issues to be resolved, with transparency, are ethical conduct, legal compliance, and accountability to stakeholders (Figure 1.7). Transparency means that the issues, factors, and outcomes of decisions and actions are open for communication, and therefore scrutiny. Ethics and accountability are concerned with decisions made with

---

- Decisions are made in consideration of the expectations of society.
- Accountability, transparency, ethics, and stakeholders should be factors in the decision-making process.

FIGURE 1.7   ISO 26000 issues of corporate governance.

fairness and equity. If decisions are made ethically and transparently, then legal compliance is more likely.

Remember, transparent and ethical behavior is part of the definition of social responsibility. Accountability, transparency, ethical behavior, respect for stakeholder interests, respect for the rule of law, respect for international norms of behavior, and respect for human rights are the principles of behavior across all ISO 26000 core subjects. These concepts are reiterated throughout the guideline. The core subject of corporate governance transcends the other six core subjects. It should be integrated through all aspects of an organization's social responsibility program.

One risk commonly associated with corporate governance is the lack of long-term economic sustainability of the organization itself. Decisions made for short-term gain, in secret, or for the benefit of some in the organization at the expense of harm to others are socially irresponsible behaviors in the core subject of organizational governance. Failures caused by irresponsible corporate governance can include the loss of leadership due to jail or expulsion, the loss of stock value or shareholder financial support, the failure to attract the best human capital due to prejudiced or unfair employment practices, and unpredictable decision-making processes that suboptimize outcomes. If systems to ensure appropriate organizational governance are not in place, focusing on improvement in the remaining core subjects may not be effective. Unethical behaviors could be occurring, undetected, making a mockery out of the organization's whole social responsibility initiative. Organizational governance must be the starting point of the improvement of social responsibility performance.

1.5.2 HUMAN RIGHTS

ISO 26000 pulls from the International Bill of Human Rights in order to define this important social responsibility subject. The right of all humans to fair treatment and the elimination of discrimination, torture, and human exploitation are the goals. Upholding value and respect for all humans is the basis of this core subject. And it is important to note that this core subject and expectations of socially responsible human rights often transcend legal compliance.

ISO 26000 defines eight different issues surrounding human rights: due diligence; risk situations; avoidance of complicity; discrimination; civil and political rights; resolving grievances; economic, social, and cultural rights; and rights at work (Figure 1.8). Examples of these issues might include purchasing

- Due diligence
- Risk
- Complicity
- Discrimination
- Resolving grievances
- Civil and political rights
- Economic, social, and cultural rights
- Rights at work

FIGURE 1.8   ISO 26000 issues of human rights.

practices that do not consider countries of origin known for violations of political rights, employment practices that are complicit to human trafficking, or discriminating against the disabled. These are very broad issues, and in many countries the lack of social responsibility toward human rights may be egregious and explicitly condoned. ISO 26000 speaks of the socially responsible organization attempting to increase its "sphere of influence" on human rights. It recognizes, for example, that eliminating supply chain partners in China in order to avoid complicity with human rights abuses may not be immediately reasonable. Leveraging commerce in China in order to improve the states' awareness and responsibility toward human rights may be a more socially responsible improvement.

Examples of failures of social responsibility toward human rights might be a failure to test for discrimination through hiring practices, a failure to ensure that labor sources are not creating demand for human trafficking, or supply chain sponsorship of child labor or compulsory labor. In short, the risks against social responsibility are primarily those that exploit the humans in the organizational system. Care for how stakeholders and their families are treated with respect to their rights of fairness as humans will improve this important social responsibility subject.

1.5.3 LABOR PRACTICES

Labor practices, as a core subject in ISO 26000, are primarily interested in employment policies and working conditions, based on the International Labor Organization of the United Nations. This builds on the aforementioned subject of human rights, but specifically addresses the responsibility of the organization as employer. It is important to note that these responsibilities as employer extend beyond just the direct relationship with the employee himself or herself, but also to the families of the employee, and the local community.

There are five issues of concern with respect to labor practices: employment relations, work conditions, social dialogue, health and safety, and human development (Figure 1.9). These issues deal with the building of sustainable social systems defining the interaction between the employer and the employee. Many countries have laws on minimum expectations around employment relations, work conditions, and employee health and safety. But social dialogue and human

- Employment relations
- Work conditions
- Social dialogue
- Health and safety
- Human development

FIGURE 1.9   ISO 26000 issues of labor practices.

development may be less recognized labor practice issues. For example, ISO 26000 details social dialogue as the collaboration between employee and employer in solving business problems, such as plant closings or plant openings. Having social dialogue may create the expectation that both employee and employer have mechanisms to solve business issues that engage both parties. For example, a socially responsible organization with social dialogue may request a task team, which includes the employee stakeholders, to evaluate alternatives to a plant closing prior to making the decision.

Failures of socially responsible labor practices would be indicated by the lack of recognition of employees as key stakeholders in the organization's sustainability. Exclusion, conflict, or disregard of the benefits of employees as important factors of sustainable social systems is a risk to social responsibility. Recognizing employees as embedded elements of the society side of stakeholders will improve the issue of labor practices as an element of social responsibility.

1.5.4 ENVIRONMENT

Environmental issues are, many times, the sole focus of sustainability programs. However, it is important to note that it is the norms of the society in which the organization operates that influence sensitivity to environmental impact. For ISO 26000, environmental issues are important, but not the only core issue of social responsibility. In general, ISO 26000 approaches environmental sustainability through life cycle management and efficiency. Organizations should recognize the cradle-to-grave goal of product and service delivery. What happens to a product after the consumer has extracted his or her desired value is an important aspect of the product. Also, minimizing wasted natural resources is an issue of this core subject.

There are four issues identified by ISO 26000 with respect to environmental sustainability: prevention of pollution, sustainable resource use, climate change mitigation and adaptation, and protection and restoration of the natural environment (Figure 1.10). For example, regardless of the agreed upon causes of climate change, or even agreeing upon its existence, if it happens, climate change is a physical risk to organizations. With the possibility of rising sea levels, an organization investing heavily in coastal areas may present a risk of social responsibility failure. Also, for example, sustainable resource use should be planned throughout the life cycle of a product. A vehicle manufacturer who is focused

---

- Prevention of pollution
- Sustainable resource use
- Climate change mitigation and adaptation
- Protection and restoration of the natural environment

FIGURE 1.10   ISO 26000 issues of the environment.

on building high-efficiency electric vehicles with no focus on energy efficiency in its manufacturing processes is only meeting half of its social responsibility. A vehicle manufacturer who builds high-efficiency vehicles, in a highly efficient factory, with no regard for the energy required to recycle the vehicle material after it has been scrapped is also not meeting its social responsibility expectations.

A failure to recognize environmental impacts for the full life cycle of products and services is the risk for this issue of social responsibility. Sustaining an abundant ecology for future generations is the goal. A minimization of resource waste and the reduction of impacts will improve the issue of the environment as an element of social responsibility.

1.5.5   FAIR OPERATING PRACTICES

The concept of fair operating practices is concerned with building systems of fair competition. Preventing corruption, encouraging fair competition, and promoting the reliability of fair business practices helps to build sustainable social systems. An organization that avoids the responsibility of fair competitive practices may disrupt the ability for any business to succeed in that environment. The core subject of fair operating practices is interested in how organizations treat each other.

Anticorruption, responsible political involvement, fair competition, using a responsible sphere of influence, and respect for property rights are the five important issues of fair operating practices identified by ISO 26000 (Figure 1.11). An organization that bribes political officials in order to influence legislature that may harm competitors would not be operating in a socially responsible manner. An organization that is involved in the political process to ensure fair and equitable laws, taxes, or oversight is helping to build a sustainable society. A government that deploys a burdensome tax, thus preventing fair profits, is acting irresponsibly. An NGO that breaks laws in the registration of voters causes irresponsible political involvement.

A failure to achieve responsibility with respect to fair operating practices may benefit the organization in the short term. However, long-term business is built upon mutual trust, trust between customers, suppliers, and employees. An intention to operate unfairly prevents the development and maintenance of this required trust. A sustainable society is one in which fair and equitable opportunities are available for all organizations.

- Anticorruption
- Responsible political involvement
- Fair competition
- Using a responsible sphere of influence
- Respect for property rights

FIGURE 1.11   ISO 26000 issues of fair operating practices.

- Fair marketing, information, and contracts
- Protecting consumer health and safety
- Sustainable consumption
- Consumer support
- Consumer data protection and privacy
- Access to essential services
- Education and awareness

FIGURE 1.12 ISO 26000 issues of the consumer.

1.5.6 CONSUMER ISSUES

Consumer issues are primarily concerned with product safety and quality. This issue of social responsibility is concerned with the protection of the consumer and his or her rights. Organizations need the trust of the customer in order to perpetuate transactions with those customers. Protecting the health and welfare of consumers helps to build that trust.

ISO 26000 guidelines toward consumer issues are based upon the United Nations Guidelines for Consumer Protection. There are seven important issues: fair marketing, information, and contracts; protection of consumer safety and health; sustainable consumption; consumer support; consumer data protection and privacy; access to essential services; and education and awareness (Figure 1.12). This comprises myriad interrelated issues. Socially responsible consumer issues include honest and thorough communication between the provider and the consumer, which includes a concern for the whole consumer experience, including the information shared in that experience.

Failure to achieve this responsibility may risk the life of consumers, including their children, and future generations. This responsibility is not to be taken lightly. The impact of the product or service, from the development of the product to the disposal of the product, should be safe and sustainable for the consumer. Treating the consumer's experience as an afterthought is a failure of social responsibility. For example, targeting lesser-educated consumers for subprime home mortgages is a failure within the social responsibility core subject of consumer issues. Failing to secure customer privacy information is another consumer issue. Protecting the consumer from an adverse impact caused by the provided product or service is socially responsible.

1.5.7 COMMUNITY INVOLVEMENT AND DEVELOPMENT

Society exists in local communities. Communities linked together create societies. Societies linked together create the interrelated world in which we live. The core subject of community involvement and development intends to create sustainable environments where increasing levels of education and well-being can exist. Stating community involvement as a core subject raises the awareness that actionable social sustainability starts in the local community.

There are seven core issues of community involvement and development expressed in ISO 26000: community involvement, social investment, employment creation,

- Community involvement
- Social investment
- Employment creation
- Technology development
- Wealth and income
- Education and culture
- Health
- Responsible investment

FIGURE 1.13   ISO 26000 issues of community involvement and development.

technology development, wealth and income, education and culture, and health (Figure 1.13). Creating healthy communities ensures a sustainable society. A focus on the needs and benefits of the community is socially responsible. It can be overwhelming for many organizations to consider improving whole societies. However, a focus on the development and involvement within the local community in which the organization operates is more easily approached.

A failure to consider community involvement is a failure to build a sustainable society. Therefore, social sustainability begins with community development and involvement. Just as we started with organizational governance as an overarching principle that is embedded in each of the remaining core subjects, community involvement is the foundational subject upon which to build social and ecological sustainability. Community involvement is the means by which social responsibility is deployed.

1.6   THE PREPARATION FOR SOCIAL RESPONSIBILITY PERFORMANCE IMPROVEMENT

So, now you see. We could not develop a tool for improving social responsibility performance until there was a globally recognized definition, strategies for decisions and actions, principles of behavior, and standard reporting mechanisms. But upon the foundation of these important guidelines, combined with our experience in product and process quality improvement, it becomes apparent to us that these old, well-worn, quality tools are ready for this new topic of social responsibility. We sought the ideal of quality and, through the concept of continuous improvement, found a path. Now, while seeking the ideal of social responsibility, the same path lies ahead.

But now what? What do we do after we become aware of the risk? Let us teach you how to use one of these tools toward the mitigation of risks found within those seven core subjects of ISO 26000. The ISO guideline has shown us to be aware of these issues. Through the use of failure mode effects and analysis, we can show you how to identify, in detail, the severity and occurrence of negative impacts, prioritize those impacts, and take corrective action through problem solving and continuous improvement on the risk of the negative outcomes, before the irresponsible action has ever taken place.

2 Social Responsibility as a Continuous Improvement Target

What to expect:

- Social responsibility as a tool for action
- Social responsibility as a process-based approach
- Social responsibility as a business imperative
- Risk abatement of social responsibility
- Risk assessment of social responsibility

2.1 A TOOL FOR ACTION

This book has been written to fill a void. We have found many books, articles, and resources focused on increasing the awareness of social responsibility. There are guidelines, standards, awards, Web sites, and conferences dedicated to the display of the best socially responsible practices. However, there is very little information in the way of *how* to become a more socially responsible organization. When it comes down to tactics, what to do, how to move to action at all levels of the organization, the body of literature comes up short. The context of this book is to fill that gap. Our goal is to give you tools to *become*—not just talk about becoming—socially responsible.

As we explained in Chapter 1, our definition of social responsibility resides with the International Organization for Standardization: ISO 26000. Many volunteers, across the globe, have been, and still are, working tirelessly to create this international guideline of social responsibility. We feel that strict adherence to the ISO definition of social responsibility ensures a common language, with standard definitions, and helps to build quality and validity to claims of improving social responsibility. We have reviewed the guidelines for Leadership in Energy and Environmental Design (LEEDS) certification measuring and building sustainability, magazine awards for best places to work, such as Fortune and Barron's, the U.S. government's Environmental Protection Agency guidelines, thousands of community awards, and many business schools curricula. All are noble and mindful definitions of the various aspects of social responsibility. However, we feel that the coming of ISO 26000 provides a much needed global, foundational definition of social responsibility.

In Chapter 1 we also explored the ISO definition of social responsibility as "the responsibility of an organization for the impacts of its decisions and activities on society and the environment, through transparent and ethical behavior that contributes to sustainable development, health, and the welfare of society." There are some important additional concepts to comprehend within this definition. First, it applies to any organization, not just corporations. We have moved beyond corporate social responsibility (CSR) to just SR. This is the recognition that any organization, not just corporations, can be socially responsible. Schools, governments, not-for-profit organizations, clubs, and households can all be socially responsible; we do not require "incorporation" to focus on our social responsibilities. And the methods that we are describing in this book can be used just as easily toward large corporate systems as they can be to improve the socially responsible behavior in your household.

Second, social responsibility includes society and the environment. This differs from the common definition of *sustainability*. Usually the term *sustainability* is used in the context of environmental sustainability, which is typically focused on the preservation of natural resources and habitat. Surely environmental sustainability is a subset of social responsibility. But with social responsibility we also consider elements that preserve the social resources and social habitats in which we live. In the context of the ISO 26000 definition of social responsibility, sustainable social *and* ecological systems are desired; sustainability is not relegated to ecological conservation.

Third, this definition focuses on transparent and ethical decisions and actions. This is a tactical aspect of the definition. ISO's definition of social responsibility speaks not only to desired outcomes but also about ways in which to behave and act. ISO's definition recognizes that social responsibility is a process of decision making and action planning. The very definition of social responsibility, the ISO 26000 definition, recognizes that decisions and actions are critical; this is a behavioral focus, and a certain type of behavior—transparent and ethical behavior—is required when completing decisions and actions. The way in which the ISO working group develops and creates the definitional material is shown in Figure 2.1. Ultimately, the goal for this book is to give you a tool to improve the social responsibility of your decisions and actions. Becoming aware of the subject and issues is only a beginning. Without tools to drive the behavior of decisions and actions social responsibility cannot be achieved.

The achievement of an internationally recognized definition of social responsibility, by such a well-respected organization as the International Organization for Standardization, is a critical first step in the ability to approach social responsibility as a continuous improvement objective in an organization. Without a standard definition, there is the risk of not fully understanding the complexity of the issues being addressed. Without a standard global definition, social responsibility behavior in Thailand does not need to achieve the same outcomes as that in Germany. Without a standard definition, we cannot reliably measure our current state against a baseline expectation in order to measure improvement over time. Without a standard

1. The ISO Technical Board Working Group for Social Responsibility prepares a draft of the material under development (input from a cross-functional group including all stakeholders is solicited),
2. The draft is sent out to the member bodies of ISO for voting.
3. If 75% of the member bodies approve the material it becomes a part of the approved standard.

FIGURE 2.1 How the International Organization for Standardization works.

definition, best practices cannot be benchmarked. A common, global definition of social responsibility, and guidelines for the subjects and issues involved in achieving social responsibility, is the necessary beginning of continuous improvement.

2.2 A PROCESS-BASED APPROACH

We cannot stop advertisers from attaching the "green" label on every other product. Consumers will become jaded, and hopefully better informed, and the label will eventually lose its meaning. We don't want to stop CEOs and professional organizations from motivating and rewarding responsible outcomes. But setting ideals for outcomes doesn't provide guidance on how to achieve those outcomes. Simply becoming aware or informed on the desired outcome is a start to achieving the outcome, but more is needed. We want to provide our readers with practical, applicable tools, well-practiced tools from the respected reliability and quality engineering profession, by which to make decisions and take action toward improving all aspects of the social responsibility performance of their organization.

We have chosen a process-based approach toward achieving socially responsible behavior (Figure 2.2). The beginning, the middle, and the end of the process must be considered. The beginning is the definition, the awareness, and the information. Just knowing, through a clear, internationally recognized guideline, what expectations exist from socially responsible behavior is the beginning of process improvement. The tools we will be teaching are the middle of the process-based approach. How the risk of behaving irresponsibly is assessed and prioritized, and what corrective actions are taken to mitigate the risk, is the heart of the process. However, with a process-based approach to social responsibility, the end state is a little more difficult to conceive. There is no such thing as having achieved perfect social responsibility. The minute we've accomplished one level of performance, another level is sought. This then creates a closed-loop process. The end of the process-based approach to achieving social responsibility is a return to the beginning.

Social responsibility performance improvement is a process, an ideal, not an end state. The minute one improvement is made, another is possible. We need tools to deploy in that journey, tools that work at all levels of progress, on all aspects, on all processes. Risk mitigation through failure analysis is one of those tools. This book

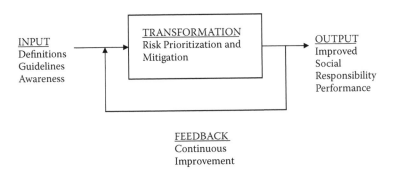

FIGURE 2.2 A process-based approach.

differs from many others we've seen on social responsibility. It is intended to teach methods of prioritization, decision making, and action by anyone in the organization. We accomplish this by defining and instructing a process. A process is very different than a goal. We could have written a book that prescribes specific actions, such as eliminating all uses of chlorine bleach from your organization. However, that prescription may not be the most socially responsible action for your organization right now. Chlorine bleach, even though in quantities it is a known pollutant, is still one of the best defenses against some of the most aggressive bacterial infections in hospitals. It would be socially irresponsible for hospitals to ban the use of chlorine bleach; this would result in undue suffering and death from infection. Offering prescriptive outcomes toward the aspects of social responsibility does not improve social responsibility in the long term; in fact, it can result in the opposite. A process-based approach, which requires the investigation of all aspects, inputs, and outputs unique to the system that is the target of improvement, is a better approach.

Teaching people how to assess and prioritize the complex aspects of social responsibility—in other words, teaching people how to *think* more socially responsible—will result in gradual, careful, mindful improvements. Our advice on the use of failure mode effects and analysis is intended to teach any person in your organization how to think, how to assess, how to decide, how to prioritize, and how to take action, on your organization's unique processes and problems (Chapter 4 details this process). Your prescriptions for improving social responsibility will be different than anyone else's. They should be. Your processes are different than anyone else's. By following our method of analysis and improvement you will find creative solutions ideally suited for your organization at this moment in time.

That's yet another complexity in improving social responsibility performance. The standards change over time. As science and technology advance, what we once thought benign now becomes toxic. What we once thought could benefit from deregulation becomes anarchistic. What we once thought was fair we now see results in inequity. Norms and standards change over time. This is another reason to avoid prescriptions for socially responsible outcomes, and yet another reason why teaching people how to think and behave toward improving social responsibility is a better approach. As you complete your first rounds of failure analysis and risk mitigation on selected processes in your organization through the use of the following methodology, you should recognize that these actions are only applicable today. Tomorrow there will be new information that will initiate the understanding of new risks and hence new actions. Having a process of improvement is much more important than having prescribed outcomes.

2.3   TRENDS IN SOCIAL RESPONSIBILITY

2.3.1   SCHOLARS AND PRACTITIONERS

We believe there are two contradictory and significant practitioner contributions to the business imperative surrounding social responsibility. There is Milton Friedman's perspective and that of W. Edwards Deming. These are two philosophically opposed giants in our history of business management, and yet they both have points of view to the application of social responsibility.

Milton Friedman in 1970 wrote an article for the *New York Times Magazine* that sternly denounced businessmen for claiming that businesses had a social responsibility in addition to their financial responsibility. Friedman's premise was that only individuals can have responsibilities; therefore, to claim that a business is socially responsible is an artificial claim. A manager of a firm is merely a representative of the stakeholders in the firm and has the responsibility to assign and utilize resources as needed to lead the firm toward profitability. If a manager decides to use the resources of the company for social projects, Friedman questions whether there is agreement by the key stakeholders for the money of the firm to be utilized in this fashion.

In Friedman's view, an organizational program of social responsibility would be a fraud. His suspicious view of firms using the veil of social responsibility to bolster favor with stakeholders sees corporate social action as inappropriate, at best, and fraudulent, at worst.

We believe there is some credence to this perspective. In some cases today, we see large corporations writing checks with lots of "zeroes" to charitable organizations without any tangible involvement (other than the monetary contribution) in the support of the social issue for which the charity addresses. For example, a large corporation may send a million dollar check to a famine relief charity in an unstable country in Africa, and at the same time be supporting the violence causing the social instability and famine through the purchase of minerals mined by the organized criminals responsible for the violence. This is not socially responsible. In this example, we question the intention behind the philanthropic gift. We question the genuine interest in the social sustainability of the receiving community.

However, we disagree with Friedman's conclusion that only individuals can behave in a socially responsible manner. Clearly an organization would desire that all individuals operate with a sense of social responsibility. But their collective work, the organization of the individual work, can also be done in a socially responsible (or socially irresponsible) manner. Every employee of the company may take great care, individually, to use only recycled products in his or her home and office, and yet if the company delegates the purchase of nonrecyclable coffee cups to an irresponsible caterer for their annual stockholder meeting, the organization has made a socially irresponsible action, while the employees are individually operating in a socially responsible fashion.

Deming, on the other hand, believed that firms could improve quality while simultaneously reducing costs via elimination of waste, rework, attrition, and litigation. Deming's perspective finds that seeking the ideal of quality ultimately leads to improved profits. Deming supports a social agenda in those cases where the result from socially responsible efforts leads to improved quality and productivity of the firm. Deming recognized that there is a "profound knowledge" required by managers (refer back to Figure 1.3). This profound knowledge acknowledges four integral factors in every system: the appreciation of the system, the knowledge of variation, the theory of knowledge, and the knowledge of people. He mandated that anyone in a management capacity should be constantly aware of these organizational factors.

We believe that Deming's profound knowledge can be easily applied to the social responsibility endeavors of an organization. In order to change the behavior of the organization, toward one that is more cognizant, and striving for continuous improvement of its social responsibilities, leaders must use profound knowledge. We must understand how human rights interact with corporate governance, and how corporate governance interacts with environmental protection. We must appreciate that a systems view is necessary. We must realize that there is variation in this system—that things can go wrong and behaviors can be irresponsible. We must plan for this variation and be proactive against inadvertent outcomes. We must have an epistemology of social responsibility. This should not be based on the whims of the individual leader, or the buzzword of the moment. There should be a theory of knowledge of social responsibility, and we must recognize that behaving in a socially responsible manner is ultimately based upon the behavior of people. We must understand the psychology of the humans in the system in order to affect their behavior. Deming's profound knowledge can be the foundation for a program of continuous improvement of socially responsible behavior in an organization.

As applied to social responsibility, the Friedman epistemology of the organization is one that is focused on the individual's behavior. No doubt, individuals with no regard toward ethical behavior can be a problem for the socially responsible organization, and an organization made up of socially responsible individuals is a great starting point. But Deming teaches us that the whole system is important, and if we forget the interactions of system variation, foundational knowledge of the system, and the people in the system, we won't get very far in changing the outcome of those organizational systems. Just like Friedman, we don't want just corporate philanthropy, and just like Deming, we do want to understand the whole system before we deem one action more responsible than another.

2.3.2 ACADEMIC RESEARCH

Social responsibility as a topic of business management is still in its infancy. There is still a considerable amount of focus merely on defining and categorizing what is socially responsible. The characterization of relationships between social responsibility and other constructs is also fairly new and under development. The most common relationship researched is the impact of social responsible initiatives on organizational performance. Deming's viewpoint, that businesses can improve quality while simultaneously reducing costs, is supported in academic research with findings that suggest socially responsible efforts result in improved business performance.

Contrary to the Freidman viewpoint, and just like improvements in product quality, improvements in social responsibility can lead to a competitive advantage, improve brand image, and result in innovative new products. For example, the Prius established Toyota as the market leader in hybrid design and technology, with the first mass-produced hybrid vehicle. Ben & Jerry's socially responsible approach to their ice cream business promoted social activism equal to economic performance. Terracycle recently joined with OfficeMax to develop ecologically improved office and school supplies; this is in addition to their well-known innovative "worm poop" gardening products and reuse of plastic bottles. These social responsibility brands and products have demonstrated market superiority through responsible products.

Research that suggests socially responsible efforts result in improved business performance is also emerging. For example, researchers have found that social responsibility actions affect consumer decisions only when the motives of those actions are credible. An additional study found the nuance that companies that focus on interpersonal dimensions in their social responsibility programs achieve greater financial performance. Most of the extant research finds that the business imperative of social responsibility is customer loyalty. The market appears to be loyal to those products that make us feel as if we are contributing to social and ecological sustainability.

And we shouldn't forget the responsibility of the consumer to have the intention of social responsibility. After all, it is the behavior of the customer that will influence the behavior of the business. Research has found, in consumer behavior that has influence on the corporation, recycling, and usage reduction are important motivators. The burgeoning scientific research on the benefits of a business to improve social responsibility performance, and the factors of importance to the socially responsible consumer, is beginning to show the same types of behaviors we've seen from the quality movement. Not only are companies listening to their customers, but the customers are listening to the companies. And everyone's expectations are increasing.

Improvements in social responsibility can lead to improvements in competitive advantage and brand image, lead to innovative new products, and may be necessary for customer loyalty. This awakening recognition of the need for social responsibility as a competitive business management practice is creating pressure for the development of the international guideline. There is demand for a cohesive foundation upon which to build organizational social responsibility performance improvement programs.

2.3.3   THE BUSINESS IMPERATIVE

In the spirit of transparency, it wouldn't be fair for us to leave the chapter on the context of the book without discussing the business imperative of social responsibility. As stated earlier, it is our belief that social responsibility is following a path similar to that of the quality movement. Since the broad dictums of total quality in the late 1970s, the world has experienced the market imperative of product and process quality improvement. Attention to organizational decisions and actions that yielded improvements to quality products, gradually, persistently, over time, has been the claim to product quality fame. It is our prediction that social responsibility will follow the same path of maturation as the quality movement.

As we see now, there are many slogans and loose definitions, broadcasts and awards about social responsibility, just as there was in the late 1960s and early 1970s in the quality movement. There are a few technically adept practitioners, but no methodologies broadly disbursed. As with the publication of ISO 9000, ISO's standard for quality management systems, the publication of ISO 26000 will create common definitions and guidelines for organizations. Businesses will begin to align with the ISO guideline. Awards, marketing, and advertisements will begin to align with the ISO guideline. More practitioners will be trained. Credibility for good practices will be achieved, and poor practices will wash out. There will always be those organizations dedicated only to slogans, as there are marginal organizations claiming ISO 9000 compliance in quality today. We believe, just as with quality, those organizations with a deep understanding and abidance to social responsibility will achieve superiority in the marketplace. A competitive advantage today will become a minimum expectation in the future. Social responsibility, like quality, is about doing the right things the right way for the long-term benefit. Those organizations dedicated to this ideal will succeed in the end.

2.3.4 ACCELERATION OF CONCERNS

As firms expand globally they must manage multiple stakeholders in multiple countries, cultures, and political and business environments. The global marketplace has made consumers and nongovernmental organizations more susceptible to the socially irresponsible behaviors of organizations. Because of the increasing complexity of the supply chain as organizations move toward globalization, it is more difficult for organizations to know the supplier of their supplier. This complexity of supply networks has exposed the vulnerabilities of organizations and is driving the need for improved social responsibility performance. In the following sections we will discuss some examples of irresponsible behaviors that are driving this interest in social responsibility.

2.3.4.1 Recent Debacles in China With Product Liability

There appear to be an increasing number of concerns regarding social responsibility as businesses are increasing their supply chain presence in China. Several issues dealing with product liability have hit the media headlines. One of these issues is the excessive amount of lead paint in children's toys. Fisher-Price (Mattel) announced the recall of 83 types of toys, including the popular Dora the Explorer and Elmo toys in 2007. With almost 80% (according to Canwest News Service) of the world's toys being manufactured in China, this is a key area of concern. Any toy for children, according to U.S. Federal Regulations, cannot have more than .06% lead. Higher contents of lead can lead to toxicity. Here is an example where Mattel's attention to social responsibility should have expanded beyond its four walls to its suppliers to ensure that the ultimate stakeholders, the child consumers, were not negatively impacted.

Child labor concerns were also brought to the forefront as companies began to expand into China. UNICEF estimates over 200 million child laborers in Asia (USA Today, September 3, 2007). Gap, Inc. had to address concerns by the Save Childhood Movement regarding the use of children under 10 that were being housed in a supplier facility in India. The children were not free to come and go. This housing of young workers is akin to modern-day slavery. Companies that operate in countries that have been reported for labor exploitation risk should take the necessary steps to ensure that their products are produced in a socially responsible manner.

Child labor concerns were also brought to the forefront as Mattel had to grapple with an over 20 million unit recall of toys made in China. The National Labor Committee, later in 2007, uncovered forced labor conditions of up to 90 hours a week and pay as low as 46 cents per hour in the factory that was producing toys for

Mattel. These investigations identified other corporations that were utilizing these factories under similar working conditions. Companies in most cases know where their products are being produced and by whom. It is the responsibility of the purchasing firm to understand and be active in mitigating social risks.

Other instances of modern-day slavery are being uncovered and reported. In May 2008, a *New York Times* article reported of the exposure of a child labor ring. Children between the ages of 13 and 15 were rescued by Chinese authorities. These children were working in Dongguan, one of the largest manufacturing centers for electronics and consumer goods. The rescued children told stories of being kidnapped and tricked by employment agencies. These agencies subsequently sent them to the manufacturing facilities in Dongguan. Once at the facilities, they were forced to work over 300 hours per month according to government officials. Interestingly, the corporations involved in producing products at these facilities were not named.

Each case of a product liability debacle or the reporting of child labor abuse demonstrates the need for the corporation involved to assess, investigate, and mitigate the potential social responsibility risks. The costly impact to profitability, brand image, and harm to human life could have been avoided.

2.3.4.2 Food Shortages

There is evidence that recent food shortages are due in part to the changing climate in various parts of the world. The areas most affected by these food shortages are the poor, who rely on agriculture as their primary means of survival. A group of Stanford researchers led by David Lobell and Marshall Burke has identified two hot spots: South Asia and southern Africa. They have deemed these areas most likely to be affected by food shortages due to climate change. They have identified corn in southern Africa as being negatively affected by hotter temperatures and lack of water. Their recommendation is to prepare these areas with training on the planting of alternative crops or the development of more heat-resistant versions of existing crops. While the findings for southern Africa only identified corn and wheat as seeing a decline, their simulation findings show a decrease in all of the major crops in South Asia as being affected by the hotter temperatures and lack of water.

There may even be causes of food shortages beyond climate change. Food shortage, in the future, may be caused by the emphasis on growing crops to make ethanol, versus growing crops to feed humans and farm animals. They caution that limits being placed on oil and flour purchases, in some areas, are impacting the supply of rice, which is a staple item for the poor in Asian countries. Restrictions on the use of dwindling fossil fuel reserves will result in an increase in energy prices, therefore increasing living and operating costs.

The potential violence and social unrest caused by food shortages is an impetus for firms to give consideration to the influence that they potentially play in this serious issue. Concern for business sustainability alone should hasten the need to improve this aspect of social responsibility. Regardless of the cause, the risk of climate change is a clear and present danger. How an organization chooses to mitigate this risk, as well as its involvement toward reducing it potential contribution to the problem, requires socially responsible behavior.

2.3.4.3   Breaches of Ethical Leadership

The core subjects of human rights, consumer issues, and environmental impact due to climate change are ready examples of the need for social responsibility performance improvement. The newspaper headlines are also filled with the need for change with respect to some of the ISO 26000 principles. Ethical leadership failures, when they are found, quickly make it to those headlines. There is a belief that the moral integrity of a firm's leadership determines the ethical culture of an organization. It is a shame to say that there are many past (Enron, Tyco, and Vivendi) and present (Madoff, Comverse, and HP) examples of unethical leadership within organizations. One of the most recent examples of unethical behavior is that of Bernie Madoff. Bernie Madoff was a trusted financial investor to many high-visibility personalities, nonprofit organizations, and firms. Over a period of many years he used the investments of new clients to fund above-average returns for the early investors within his firm. His exclusive Ponzi scheme is said to have been valued at $65 billion.

Leaders of organizations, you would think, should be aware of the publicity damage inherent in these responsibility breaches. Three examples of unethical practices can be found for Wal-Mart and Brody Records. Each of these companies has been in the news for utilizing Internet blogging in unethical ways. Wal-Mart initiated a campaign that was supposedly following a family traveling across America and stopping at Wal-Mart stores along the way. Many people tuned in to follow the family and became personally interested in their journey. However, it was uncovered that this was not a real family, but merely a marketing campaign by Wal-Mart.

Another case involves Brody Ruckus, who on Facebook claimed that if he could get 100,000 people to join his group, his girlfriend would engage in certain sexual activities with him. If he could get 300,000 people to join, then these people would have access to pictures of the event. It turns out that Brody Ruckus was not a real person, but the interest developed around his project to get people to join his group resulted in Brody Records/Ruckus Music Downloading's access to 300,000 e-mail addresses, which it could utilize to market its music downloads.

It is easy to find many cases of firms that have been fined by the Federal Trade Commission (FTC) for violating children's privacy. Xanga, Hershey Foods, and Mrs. Field's Cookies were all fined for requesting personal information from children and teens online without their parent's consent. One would think that leaders of these various organizations are aware or knowledgeable of the Internet activities that are ongoing with regard to their organization. Any leader who claims lack of knowledge of the unethical practices active within his or her responsibility is demonstrating a lack of control over the culture of the organization to allow these types of practices to occur.

There are many examples of business success through social responsibility and business failure due to social irresponsibility. The emerging international guideline is needed to steer organizations in the direction of managing in a more socially responsible way. The growing awareness of these failures is beneficial to the social responsibility movement. However, we need to rapidly move beyond guidelines and case studies and demonstrate actionable tools for achieving success through improved social responsibility. The context of this book is to deliver a process of improvement. It is to teach anyone in your organization how to become aware of risks against social responsibility for your processes and products. This book will teach you how to prioritize those risks to discern the most critical in need of improvement, and it will teach you how to take action on those risks, to mitigate them, for now, in the journey of seeking social responsibility performance. All of this will be accomplished through a focus on the seven core subjects and the seven principles defined in ISO 26000.

2.3.5 Conclusion of Concerns

We recognize the competitive advantage of being socially responsible and recognize the need for guidelines of socially responsible behavior, but neither is a guarantee that organizations will behave in this fashion. We also need to recognize that the improvement of social responsibility performance is a journey, and at times a difficult one. International norms may conflict with respect for the law. Transparency may accelerate the need for accountability. Ethical decision making may highlight absences in the assurances of human rights. The examples provided above give us hope that there are firms that care about social responsibility and are impacting society with their proactive approaches. While there is the positive side to social responsibility, there is also the acceleration of concern.

2.4 ACTIONISTS AND THEIR ROLES

Social responsibility is a set of behaviors. Behaviors are exhibited by individuals in the organization. And improvement of socially responsible behavior requires action, in the form of behavior change, by everyone in the organization. However, the role one carries in the organization may make the role in changing behavior different. We see leaders, managers, and individuals as having different roles in the instigation of behavior change. We call the people who fulfill this instigation of behavior change actionists. In the following section we will explore several different actionist roles during the improvement of social responsibility performance. As your organization embarks on this improvement journey, it will be important that everyone in the initiative fulfill his or her unique role.

2.4.1 Leader

The leader's role in ensuring socially responsible behavior is to establish an organizational culture of socially responsible behavior (Figure 2.3). Socially irresponsible behavior should be admonished and rebuked within the organization. Socially responsible behavior should be recognized and rewarded. The leader should assess activities occurring within the organization to understand whether the actions of the representatives of the firm are socially responsible.

The leader should be involved in identifying the most critical social responsibility impacts associated with the products and services provided by the

- Awareness—Be aware of the actions and projects within the organization.
- Culture—Establish a culture that rewards socially responsible behaviors.
- Continuous improvement—Support for many, small, impactful changes must be recognized.
- Strategy—The project selected should be linked to the overall strategy of the firm.

FIGURE 2.3   The leader as actionist.

- Awareness—Be aware of the actions and projects within the organization.
- Continuous improvement mentor/coach—Provide teams with guidance as needed.
- Link to leadership—Ensure projects selected are aligned with organizational strategies.
- Project selection/control—Help the team select appropriate projects and control the scope of the project.
- Resource—Assist the team in securing the most appropriate resources for the effective completion of improvement actions.

FIGURE 2.4   The manager as actionist.

organization. Strategies should be communicated and tactical plans implemented to support the mitigation of risks associated with these critical aspects. The leader should be actively involved in reviewing the relevant risks arising from irresponsible behavior in order to integrate socially responsible elements into strategies of the firm.

The leader should encourage the continuous improvement of socially responsible actions and recognize and share best practices across the organization. The leader should be one of the biggest cheerleaders within the firm and set the example of conduct. It is critical that the leader of the organization be a role model of socially responsible behavior. All eyes are on him or her. The leader should sets expectations of continuous improvement. The expression of problems and opportunities for improvements should be encouraged. All attempts to improve, even those that do not achieve the goal, should be rewarded.

2.4.2 MANAGER

The manager of the organization plays a somewhat different role. The manager of the organization should be the coach and mentor of the teams conducting improvements (Figure 2.4). The manager should promote and encourage his or her employees to participate on social responsibility performance improvement teams. Some of the best process improvements happen with a cross-functional team; the manager should ensure that resources are available for participation. Selection of the appropriate team members for the work is another key role of the manager.

The manager becomes the conduit between the leadership of the firm and the employees. In this role the manager should help the team control focus on improving those most risky aspects of the organization. The scope of improvement projects should be aligned with the needs of the business and the strategies of the organization; the manager should oversee these aspects.

Support of the projects identified from the goals should take the form of ensuring that the best projects are implemented, that the projects are reviewed on a regular basis, and that a continuous improvement culture is instilled. The manager is the front-line observer of social responsibility. And, as with the leader, all eyes are scrutinizing the manager's behavior. It is critical that he or she model responsible behavior.

2.4.3 TEAM MEMBER

Another key player in the development of the social responsibility performance improvement initiative is the team member. The team members selected to complete improvement activities play a crucial role (Figure 2.5). It is the differing perspectives and positions within the organization that provide ideas about how each of the seven core elements of ISO 26000 could be impacted by products or services. This insight should be used to develop exhaustive failure modes for each of the seven elements given the scope of the project undertaken.

Once the risks have been identified, the team members are key resources in determining the possible losses resulting from those risks. The team should develop a team mechanism (voting, majority rule, etc.) for developing consensus on improvement actions. Once the key elements are selected, the team should certify that causes of socially irresponsible behavior have been exhaustively identified. Having the most appropriate team member take the lead role on implementation of specific actions will result in more likely success. The lead team member on the action should collaborate with the manager that has the greatest stake in the high-priority risks being mitigated or eliminated. This team member should receive mentoring and coaching from his or her manager in the successful completion of the action.

2.4.4 THE INDIVIDUAL

As a participant on an improvement team within the organization, the individual can take away learning from the analysis. These lessons can even be utilized in a nonorganizational setting, such as at home, at church, etc. For example, as found by an improvement team at work to reduce energy output, halogen lighting not only reduces energy requirments, but it also has longer life (less cost for new bulbs, less

---

- Brainstorming—Be exhaustive in identifying the potential failure modes for each social responsibility core subject and issues.
- Perspectives—Ensure that the team is made up of varying perspectives to the problem selected.
- Action leader—Select the appropriate leader for each action to ensure successful implementation.

FIGURE 2.5 The team member as actionist.

- Brainstorming—Be creative in identifying potential failure modes for each social responsibility element.
- Perspective—Be an active participant in providing your special perspective to the team.
- Ownership—Take ownership and volunteer for those action items that may fall under your direct responsibility.
- Identification—On a daily basis consider the impact of your actions and decisions on the seven core elements of social responsibility.
- Communication—Immediately communicate or add to the SRFMEA process any new failures modes identified in the course of regular business.

FIGURE 2.6 The individual as actionist.

waste to the landfill). This knowledge can be translated to reducing energy consumption and waste within the home by switching to the new halogen lighting.

The individual can also take on personal causes by joining nongovernmental organizations (NGOs) that support a socially responsible initiative that is not being addressed within his or her organization. His or her work with the NGO can be used to educate the organization on a more proactive role that it can play relative to the individual's particular concern. There are many grassroots actions that an individual can proactively take toward being more socially responsible (Figure 2.6). For example, the American Society for Quality (ASQ), an NGO, relies on volunteers to help participate in the shaping of ISO 26000.

2.5   GRASSROOTS

The ISO standard will cause a more holistic grassroots effort, we hope. There are many Web sites available for calculating your carbon footprint. As individuals, we all consume products and services, travel via air and automobile, and utilize a variety of energy sources that contribute to pollution. These calculators provide information and guidance via the questions asked on how an individual contributor can reduce his or her personal CO_2 output. Caution must be made in selecting a calculator that has elements you can monitor and control.

Many organizations, as part of their Web site, provide ways in which an individual can start a socially responsible campaign within his or her community. These Web sites typically explain how you can lobby the political representatives of your state to take action, offer specific process steps for starting an environmental campaign within your community, and provide links to organizations in which you can become involved. The identification of simple actions can be found on lists on various Web sites; for example, the 2008 Keep Australia Beautiful campaign offered the list of ten items found in Figure 2.7.

2.6   SMALL CHANGES MAKE A BIG DIFFERENCE

Many firms have taken small steps toward big differences in energy consumption, carbon footprints, and other environmental aspects. Actions such as turning off lights and computers at the end of the day, turning down the air conditioning on

1. Set thermostat at a maximum of 20°C for winter and no less than 26°C for summer.
2. Turn off TVs, computer monitors, and appliances at their power source.
3. Limit showers to four minutes.
4. Turn off water while brushing your teeth or shaving.
5. Compost or recycle your waste.
6. Say no to plastic bags; take reusable bags when shopping.
7. Use less chemicals when cleaning.
8. For floors and windows, add vinegar to every liter of water.
9. Ride your bike whenever you can.
10. Find alternatives for getting to work (carpool, public transportation, etc.).

FIGURE 2.7   Ten simple actions for social responsibility (2008 Keep Australia Beautiful).

weekend days, revising the travel policy to reduce unnecessary travel, and replacing plastic bags with reusable bags can be driven by organizations, individuals, or community groups.

In the United Arab Emirates (UAE) Carrefour, a supermarket chain, encouraged its customers to replace the use of plastic bags with reusable bags. Carrefour identified that if you use the reusable bag four times, there is a 20% reduction in greenhouse gases associated with plastic bag production and an 18% reduction in plastic bag usage. If the reusable bag is used 20 times, then the plastic bag usage is reduced by 80%. The debate on whether to use paper or plastic shopping bags, when evaluated, suggests that both products affect the environment, the paper bag in the felling of trees and the plastic bag in the production of polyethylene. The finding from these debates suggests that the reusable bag is the least damaging to the environment because it is made of recycled products and is reusable.

A local Hawaiian organization, Kanu Hawaii, encouraged residents to become more environmentally involved. The nonprofit organization believes that if you add up the activities of a small number of residents, it can have a big impact. It encourages residents to make commitments to change their behaviors and track the results. For example, if 400 residents take shorter showers, Kanu Hawaii believes that the collective efforts of resident members over a one-year period will result in enough drinking water for 2,000 people. Also, if residents spend locally, $1,603,797 will be contributed to the local economy. By encouraging Hawaiian residents to be more socially responsible, individually, the organization is able to achieve big improvements.

Behaving in a socially responsible manner can be an endeavor taken on by an individual, a corporation, a nonprofit, or anyone at anytime. Within an organization there are key roles that must be taken to ensure that socially responsible behaviors are interwoven into the fabric of the organization. Socially responsible actions should tie to the strategies and mission of the organization. Socially responsible actions should be captured, monitored, and resulting impacts documented. Organizations can also play a key role in encouraging employees, as individuals, to be socially responsible. Not-for-profit organizations also play a key role in increasing awareness and encouraging socially responsible behavior. Many small actions, when aggregated, can make a major impact to create a sustainable society and environment.

2.7   IDENTIFYING SOCIAL RESPONSIBILITY STRATEGIES

Socially responsible improvement strategies can be identified by utilizing the outputs of a risk assessment. We will be introducing one risk assessment method, the social responsibility failure mode effects and analysis (SRFMEA), in the following chapter. The outputs of a risk assessment consist of the identification of possible failures and losses associated with each of the seven ISO 26000 core social responsibility subjects. The mitigation of risks associated with these failure modes can become the nucleus of a social responsibility performance improvement strategy. For example, if the most severe failures in your organization are found to be associated with human rights violations, your strategy for the future should focus on enhancing elements to the organizational strategy that accomplish the mitigation of impacts to human rights, as opposed to actions to improve environmental protection.

Another approach for identifying social responsibility strategies is the use of traditional strategic analytic tools such as strength-weakness-opportunity-threat (SWOT) analysis to analyze the core ISO 26000 social responsibility subjects. If we stick with the human rights example, as an organization, we might find that there is a potential for human rights issues at our location in China. But, as an organization, we have a stellar record of human rights resolution in the United States. We should strategically promote this strength by ensuring that our company will have the same high human rights standards worldwide. A weakness could be identified in our ability to ensure that all suppliers in our supply chain have the same high standards for human rights. A strategic action could be to provide training or conduct assessments to ensure that all suppliers have the same standards. An opportunity might be identified if we have had competitors with human rights abuses. We can learn from our competitor's mistakes and ensure that we have a strategy for identifying and resolving any human rights issues in order to minimize the impact to the financial performance of the organization. One threat might be that a competitor has better performance than us on human rights issues, and promotes its values in its marketing materials. One strategy could be to identify the best practices for minimizing human rights issues within our industry and implementing those best practices. The identification of socially responsible failure modes can be utilized to identify concerns, prioritize actions, and incorporate elements of risk mitigation into the organization's overall business strategy.

There are many possible strategies that could be deployed in order to improve social responsibility performance. Just like many other performance criteria for organizations, social responsibility needs to have strategies for improvement identified, and tactics for improvement deployed (Figure 2.8). In this book, our focus is on using risk assessment as a strategy for identifying opportunities, and risk abatement through continuous improvement.

2.8   SOCIAL RESPONSIBILITY AND RISK ABATEMENT

In this section we will provide a brief overview of failure mode effects and analysis (FMEA) and responsibility analysis (RA). The full history and how to use them will be presented in the next chapter. Right now, it's important to understand what they

Principle	Definition of Principle	Strategies for Improvement
Accountability	An organization should be accountable for its impacts on society and the environment	1. The results of its decisions and activities, including significant consequences, even if they were unintended or unforeseen 2. The significant impacts of its decisions and actions on society and the environment
Transparency	An organization should be transparent in its decisions and activities that impact on society and the environment	1. The purpose, nature, and location of its activities 2. The manner in which its decisions are made, implemented, and reviewed, including the definition of the roles, responsibilities, accountabilities, and authorities across the different functions in the organization 3. Standards and criteria against which the organization evaluates its own performance relating to social responsibility 4. The known and likely impacts of its decisions and activities on society and the environment 5. The identity of its stakeholders and the criteria and procedures used to identify, select, and engage them
Ethical behavior	An organization should behave ethically at all times	1. Developing governance structures that help to promote ethical conduct within the organization and in its interactions with others 2. Identifying, adopting, and applying its own standards of ethical behavior appropriate to its purpose and activities but consistent with the principles outlined in this international standard 3. Encouraging and promoting the observance of its standards of ethical behavior 4. Defining and communicating the standards of ethical behavior expected from its personnel, and particularly from those that have the opportunity to significantly influence the values, culture, integrity, strategy, and operation of the organization 5. Preventing or resolving conflicts of interest throughout the organization that could otherwise lead to unethical behavior 6. Establishing oversight mechanisms and controls to monitor and enforce ethical behavior 7. Establishing mechanisms to facilitate the reporting of violations of ethical behavior without fear of reprisal 8. Recognizing and addressing situations where local laws and regulations do not exist or conflict with ethical behavior

FIGURE 2.8 Principles of social responsibility (ISO 26000).

Principle	Definition of Principle	Strategies for Improvement
Respect for stakeholders	An organization should respect, consider, and respond to the interests of its stakeholders	1. Identify its stakeholders 2. Be conscious of and respect the interests and needs of its stakeholders and respond to their expressed concerns 3. Recognize the legal rights and legitimate interests of stakeholders 4. Take into account the relative capacity of stakeholders to contact and engage the organization 5. Take into account the relation of its stakeholders' interests to the broader interests of society and to sustainable development, as well as the nature of the stakeholders' relationship with the organization 6. Consider the views of stakeholders that may be affected by a decision even if they have no formal role in the governance of the organization or are unaware of their interest in the decisions or activities of the organization
Respect for the rule of law	An organization should accept that respect for the rule of law is mandatory	1. Comply with legal and regulatory requirements in all jurisdictions in which the organization operates 2. Ensure that its relationships and activities fall within the intended and relevant legal framework 3. Comply with its own bylaws, policies, rules, and procedures and apply them fairly and impartially 4. Remain informed of all legal obligations 5. Periodically review its compliance
Respect for international norms of behavior	An organization should respect international norms of behavior, while adhering to the principle of respect for the rule of law	1. In countries where national law or its implementation does not provide for minimum environmental or social safeguards, an organization should strive to respect international norms of behavior 2. Where national law or its implementation prohibits organizations from respecting international norms of behavior, an organization should strive to respect such norms to the greatest extent possible 3. In situations of conflict with international norms of behavior, and where not following these norms would have significant consequences, an organization should, as feasible and appropriate, review the nature of its activities and relationships within that jurisdiction 4. An organization should consider legitimate opportunities and channels to seek to influence relevant organizations and authorities to remedy any such conflicts in national law and its implementation 5. An organization should avoid being complicit in another organization's activities that fail to meet international norms of behavior

FIGURE 2.8 (Continued)

mean and the intention of the analysis. Both the FMEA and RA are analytic tools. They are always performed together with the information in the FMEA leading to the completion of the RA. Ultimately there are two functions of the FMEA and RA: the identification of risk and the prioritization of the criticality of that risk to focus corrective action for the assurance of responsibility.

2.8.1 The Assessment of Risk

There are many different types of risk assessment. Each type of industry seems to have its own form. Safety engineers would be familiar with hazard analysis. Designers would be familiar with knowledge criticality analysis. The venture capitalist may be familiar with a business impact analysis. Bankers would be familiar with risk management. Insurance underwriters would be familiar with actuarial risk assessment. These types of analysis are all just different forms of risk assessment.

Risk assessment demonstrates the awareness that nothing ever goes perfectly. There is always the risk of things going wrong in both processes and products. For example, in processes people may be improperly or inadequately trained. Computer systems may crash. Data may be lost. The process may deliver the needed service too slowly. With products, for example, components wear out. Subassemblies may not meet specifications when all the variation is stacked up. Design engineers may make errors in calculations or with translating units of measure. For any of these examples, the consequence of the error is different based upon the type of error. The failure mode is the way in which the process or product fails; it describes the weak link of the chain. The failure mode effect describes what happens when the process or product failure occurs—what is the outcome of the error? The analysis of these failure modes and their effects, through FMEA, creates a ranking among all failure modes and effects against each other. The FMEA is a tool that focuses on an individual process or product and that rank orders all of the identified failures against each other.

Responsibility analysis (RA) always follows FMEA. RA takes the rank ordering of the severity of failure modes and effects and then attaches the probability of occurrence, as a rating, to the failure mode. There may be failure modes with horrible, life-threatening consequences, but protected by quadruple redundant failsafes. For example, in most modern-day aircraft there are quadruple redundant wiring schemas for control surfaces. Clearly if an airliner loses its rudder, the effect is severe. One mode by which the aircraft would lose its rudder is by a loss of electrical signal between the tail of the aircraft and the cockpit. The severity of this failure mode in the FMEA would be ranked very high.

However, most aircraft manufacturers build multiple separate systems of electrical communication between the cockpit and the tail. If one wire is severed, there may be three more to take its place. Therefore, the probability of occurrence of the loss of electrical signal between the tail and the cockpit (the failure mode) is very low. Although the single ranking of the severity of the failure mode, coming from the FMEA, is extremely high, when partnered with a very low occurrence rating, the responsibility analysis yields a lower risk priority. RA is when the severity of the failure mode and the probability of occurrence are combined in analysis to yield an overall concern for negative consequences. The RA is ultimately a prioritization matrix. We'll be going into the full detail of these prioritization mechanics later. For now, its important for you to understand that FMEA and RA happen in phases, and result in prioritizations of risks that need improvement.

For our purposes, the combination of the whole process will be referred to as social responsibility failure mode effects and analysis (SRFMEA). This includes the identification of the product or process upon which to focus, understanding the stakeholders involved in that process, creating an organizational narrative of intent toward social responsibility performance, the FMEA, the RA, and the planning of improvement actions. SRFMEA is not just risk assessment and prioritization; it is a whole process of providing the organization with a focus for its social responsibility improvements. It is a process-based approach of the continuous improvement of social responsibility performance. We are proposing the use of a risk assessment, in the form of an FMEA, partnered with a prioritization for improvement, in the form of an RA, utilizing the seven core subjects of ISO 26000, to enhance an organization's performance toward social responsibility. We are taking these well-worn reliability tools and using them on a growing concern toward social responsibility. We are taking the lessons learned from the quality movement and applying them to the social responsibility movement.

Now that we have a better understanding of the SRFMEA process, we will review why the SRFMEA as a risk assessment tool makes sense. Risk assessment is the process in a risk management program that involves putting quantitative values to the element under review. If we are concerned about environmental risk, one might want to know what is the size of the potential loss associated with the environmental risk, and what is the likelihood that the loss will occur. This quantitative approach to risk assessment allows for the identification and prioritization of all of the potential losses. This prioritization can be utilized to eliminate and mitigate those aspects of risk that are most important to the success of an organization.

A risk assessment approach makes sense when evaluating social responsibility risk because all seven of the ISO 26000 core subjects can be evaluated quantitatively. If an organizational value has been communicated such as that by Valeo, for example, that an "eco-standard" will be utilized in the design phase of all R&D projects, then Valeo will need some manner in which to assess the risk of using certain materials, suppliers, and processes in the production of its products. One way to do this would be to utilize a design failure mode and effects analysis to evaluate all aspects of a design and whether they meet the organization's eco-standard.

This risk assessment will quantitatively assess the potential failures for each element within the life cycle of the product. The FMEA will alert the organization to those aspects of risk that would have the greatest chance of failure and the likelihood that a failure would occur. Subsequent rankings of each element of risk would allow the organization to prioritize which elements of risk need to be mitigated or eliminated to ensure that it is meeting the new environmental standard. Once the selected items for improvement are identified, then action must be taken to actually mitigate or eliminate the risk. This process should happen repeatedly to ensure that the risks are addressed and subsequent high-priority elements are also addressed.

As you can see, this practice of risk assessment is a process, not a document. So when we talk about the SRFMEA, we are not talking about the document, but the process of evaluating the social responsibility risk of all seven ISO 26000 core elements, the prioritization of social responsibility subjects that have the highest risk of failure and probability of occurrence, the selection of those high-risk priorities for continuous improvement, and the process of continually improving, reprioritizing, selecting, and improving high-risk elements. This process of continuous improvement of social responsibility risk must be integrated with organizational goals and the voice of the key stakeholders.

Having specified goals on social responsibility improvement, such as reducing emissions to 10% below the recommended levels, or based on stakeholder voice, such as the regulatory requirements that emissions be reduced to a certain level, provides the boundaries of where the continuous improvements need to take place. Whether the boundary be a goal or stakeholder voice, the selection and prioritization should always tie back to the strategic initiatives and success metrics of the organization.

2.8.2   The Prioritization of Action

In the following chapters we will go into much more detail on the history of the FMEA and RA, as well as why we think they should be used with programs of social responsibility. But, before we get into such detail, we should state the ultimate goal of the SRFMEA. These are analytic tools, but the goal is not analysis. The goal is action, but action such that the organization will focus its resources on the most important social responsibility issues.

In our experience with social responsibility programs we've seen organizations focus a lot of resource on issues of minor impact. For example, we are aware of an organization that has applied significant resources, involved many in the organization, and loudly advertised its paper and cell phone recycling program. This same organization is responsible for excessive waste of nonrenewable natural gas. In this organization, there are a lot people with their hearts in the right place; clearly their management wants to do the right thing for society. We are not advocating abandoning the paper and cell phone recycling program. What we are advocating is the use of a tool that helps this organization identify its most severe impacts, and apply resources to resolve these. It's the difference between working on the most obvious (or easy) and the most important.

The use of the SRFMEA in an organization's social responsibility program documents what the organization considers the severity and occurrence of its social responsibility failures, as a snapshot in time. It provides the prioritization of focus for continuous improvement. It separates the important from the mundane, and it provides a format for the documentation of the journey of continuous improvement of socially responsible actions.

That's the ultimate goal. We want your organization to use SRFMEA for action, not for analysis. We use the analytic tools to focus our efforts. But the goal is the effort, the actions, the improvement of outcomes. What's important is not the *doing* of the SRFMEA, it's what you do with the information as a call to action.

3 SRFMEA
A Tool for Improving Social Responsibility

What to expect:

- FMEA for risk assessment
- FMEA for military systems (MIL-STD-1629)
- FMEA for automotive (AIAG)
- FMEA for social responsibility (SRFMEA)

3.1 FMEA AS ONE RISK ASSESSMENT OPTION

Failure mode effects and analysis (FMEA) is not a new concept. It is a well-worn tool commonly used during new product introduction to prepare control plans for potential product and process failures. This chapter will discuss the history of FMEA, starting as a military document for use in weaponry system designs, to the latest industry application in health care. Different historical perspectives on the process and its intentions will also be discussed. This will include controversial aspects, such as the distinction between risk assessment and criticality analysis, the difficulty of team-based completion, and common problems encountered while doing risk assessment. How this helps toward the completion of FMEA for social responsibility will conclude the chapter.

In its most basic form FMEA is a risk assessment. Ultimately, there are two factors involved in a risk assessment: assessing the magnitude of potential losses or problems and assessing the probability of the loss or problem actually happening. Risk assessment finds it roots in early statistics as the mathematical study of reliability and probability. Studies on games of chance and public studies of wealth, health, and mortality were common early subjects. Historical mathematicians such as Gauss, Pascal, and Fermat created the foundation upon which statistics, and subsequently risk analysis, was built. Risk assessment is the intersection of two important statistical factors in a system: the potentiality of loss and the probability of occurrence of that loss (Figure 3.1).

Risk assessment takes many forms, not just FMEA. Many fields of science have different methodologies of risk assessment. For example, in the United States, the Food and Drug Administration will complete a Health Impact Assessment prior to approving drugs for use in health care and treatment. Businesses may conduct a threat assessment for information technology systems and data security. Being able to effectively predict, and mitigate, risks in the fields of health care, nuclear power, and food safety is critical. The failure to mitigate risk in fields like these can

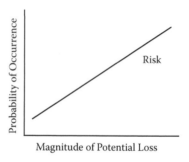

FIGURE 3.1   The factors of risk analysis.

result in the loss of many lives. But risk assessment has a robust following, even in non-life-threatening situations. For example, when deciding on investment options a business might conduct a risk-benefit assessment. To understand potential production disruptions, a manufacturer might conduct a process failure mode effects and analysis (pFMEA). Elements of risk analysis are even found in the focus groups used for marketing and advertising.

It is the FMEA foundation upon which we have built a tool to apply toward social responsibility. In many of the assessments for which the risk is immediate loss of life, the magnitude of potential is so extreme that a probability of any occurrence must be avoided. This places a heavy emphasis on detection and prediction of the risk. For the use of risk assessment toward social responsibility, there is a different goal. For this use the increased awareness of the existence of the risk is more important, and continuous improvement actions, not necessarily the elimination of the probability of occurrence, is the aim. It is for this reason that we chose FMEA, among many different risk assessment methods, as a tool for social responsibility.

Our goal for this specific type of risk assessment is to become acutely aware of the potential failure modes. The probability of occurrence is of lesser concern. Rarely will we be able to definitively assess the probability of occurrence of human trafficking, or disposing of batteries in the city waste stream. What we can do is recognize that by a weak system of assurance of legal hiring practices we may create a social failure mode that leads to human trafficking. Through failure mode analysis we can begin to recognize the risk of not having hazardous waste receptacles for spent batteries. It is less important that we calculate, with accuracy, how many batteries might be disposed of improperly without the waste receptacle. For the use of risk assessment toward improving social responsibility performance, recognizing the potentiality of loss is more important than assessing the probability of occurrence.

FMEA is one type of risk assessment that places this emphasis on the identification of many possible failure modes. It emphasizes a variety of potential losses, and therefore is a better tool for increasing the organization's awareness of many potential failure modes. As we discuss in the following chapters, FMEA allows for qualitative assessment of the probability of occurrence. Our ultimate goal of continual improvement in responsible behavior is to use FMEA to help identify many different potential issues toward which we can apply many different incrementally improving solutions. With this approach the probability of occurrence becomes an output, or result, of the continuous improvement efforts.

3.2 BACKGROUND OF THE FMEA

3.2.1 FMEA AND THE MILITARY

The FMEA's broad introduction, beyond the desk of the reliability engineer, has been attributed to the U.S. military, through the publication of MIL-STD-1629 in 1949. As a means to ensure mission accomplishment for weapon systems, this standard defined FMEA from 1949 through 1977. This specification was written with the intent that FMEA be accomplished during the design phase, as an iterative process, to mitigate product and system risk through design correction. MIL-STD-1629 espoused a two-step process: (1) All potential failure modes and their subsequent effects to mission accomplishment are documented as the FMEA, and (2) a criticality analysis (CA) is completed to better understand the probability of occurrence for each failure mode. These sequential steps resulted in the prioritization of risk, identification of corrective actions, such as the development of redundant systems to avoid single-point failure modes, and information important for the planning of weapon system maintenance actions. The FMEA analyzes the magnitude of potential loss. The CA analyzes the probability of occurrence.

MIL-STD-1629 built the expectation that risk assessment be accomplished as a team effort. Ground rules and analysis assumptions, developed by the team, are to be documented. Block diagrams and operational diagrams of the system being analyzed are the starting point. Every one on the team needs to begin with the same foundation of understanding of the system to be studied. Each of the subsequent task items is to be accomplished, by the team, as a process of analysis, not as a discrete, disjointed task (Figure 3.2).

MIL-STD-1629 created the awareness of the need of a narrative summary report; this may have been the beginning of a knowledge management system. Although both the FMEA and the CA are eventually boiled down into an indented spreadsheet, not all important assumptions, ground rules, decisions, learning points, or operational definitions can be effectively documented in this spreadsheet. The spreadsheets, often erroneously referred to as "the FMEA," were originally intended to only be a brief summary of the thoughts of the reliability engineering team. The narrative summary provides a qualitative assessment of the system and the FMEA process. It tells the story of the mission of the weapon system, the constraints on the risk assessment team, and their beginning assumptions and ground rules. The military specification recognizes that the assumptions upon which the analysis is based, the analysis itself, and the documentation are all important. The narrative summary's intent is to document the work *process*, which concluded in the FMEA and CA forms.

In the military specification the FMEA could either be completed on distinct pieces of hardware, modules, or subassemblies, or take into consideration system interactions or system function. The CA could be completed either quantitatively or

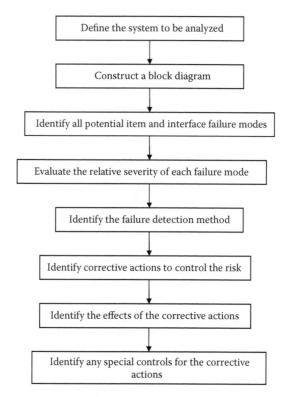

FIGURE 3.2   MIL-STD-1629 FMEA process flow.

qualitatively. Only if component reliability data were available would quantitative probability of failure mode occurrence be computed. For many criticality analyses, a qualitative analysis of failure occurrence probability is completed using five levels of probability: frequent, reasonably probable, occasional, remote, and extremely unlikely. Using a qualitative analysis of the probability of occurrence, it is not necessary to gather statistical data on the historical experience of failure. Although with mechanical systems, quantitative data are usually available from reliability testing, and thus quantitative methods would be more frequently used. However, the authors of MIL-STD-1629 recognized that sometimes these quantitative data are neither available nor needed, and allowed for the use of qualitative assessment of the probability of occurrence.

The FMEA created in MIL-STD-1629 analyzes both the failure mode, or the manner in which the weapon system fails, and the effect of that failure, or what happens in the mission as an outcome of the failure. Even though the effect of the failure is documented, the magnitude of the loss (severity) and the probability of the loss (occurrence) are calculated on the failure mode, not the effect of the failure mode. It was believed that as a design tool, the avoidance of the mode altogether would lead to the elimination of the effect. For the military standard on FMEA, the goal is to eliminate the potential of the failure mode through robust design, not the mitigation

of the effects of the failure mode through detection or redundancy. The FMEA identifies both the failure modes and the effects of the failure modes, but risk mitigation
is focused on the failure mode, thus eliminating the potential of all of the effects of
the failure mode.

3.2.2 FMEA AND THE AUTOMOTIVE INDUSTRY

This particular type of risk assessment, as a design tool for military weaponry, has
been modified and expanded in use and outcome through the years. One of the most
notable growth factors is the spread of national, international, and action group
standards. For example, ISO 9001:1987, *Model for Quality Assurance in Design,
Development, Production, Installation, and Servicing,* introduced thousands of
organizations to this heretofore obscure risk analysis method. ISO 9001 was a result
of the evolution of the British Standards Institute's BS 5750, which, ironically, was
developed from another U.S. military specification, MIL-Q-9858, originally written
in 1959. The Automotive Industry Action Group (AIAG) in the United States further
included the FMEA in it's QS 9000 standard and issued the *Potential Failure Mode
and Effects Analysis* manual in 1993 (the technical equivalent of SAE J-1739). See
Figure 3.3 for an overview of the history of FMEA standards and requirements. The
requirement for all suppliers in the U.S. automotive industry to complete FMEA on
all designs and processes rapidly spread the use and understanding of the tool, and as
what goes around, comes around, QS 9000 has now migrated back to its ISO roots,
with the migration to ISO's TS-16949 in the automotive industry.

   The evolution of the FMEA started with a military weaponry design reliability
process and eventually morphed into an element of international quality management standards. MIL-STD-1629 has been cancelled, as has QS 9000; and the use of
reliability assessment becomes more and more mainstream.

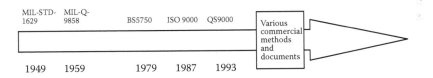

- MIL-STD-1629: "Procedure for Performing a Failure Mode, Effects, and
  Criticality Analysis," United States Department of Defense
- MIL-Q-9858: "Quality Control System Requirements," United States
  Department of Defense
- BS5750: "Guide to Quality Management and Quality Systems," British
  Standards Institution
- ISO 9000: "Quality Management Systems," International Organization of
  Standardization
- QS9000: "Quality System Requirements," Automotive Industry Action Group

FIGURE 3.3   An overview of the history of FMEA standards.

Recent use of FMEA has been spreading beyond product manufacturing. Failure mode effects and analysis and criticality analysis are now being used to successfully prevent failures in the inherently risky health care industry. The roots of FMEA and CA, as sprung from MIL-STD-1629, can now be found in tools as varied as quantitative risk analysis, reasonably available control measures, business impact analysis, Hazard Analysis and Critical Control Point System (HACCP), and knowledge criticality analysis. The fundamentals of risk assessment, analyzing the magnitude of potential loss and the probability of occurrence, are now used toward worker safety, health care patient safety, food, drug, and cosmetic quality control, software development, and environmental standards compliance. And now, we're proposing, as a polar opposite to the effectiveness of weapons of war, that FMEA and CA can be effective tools toward the achievement of social and ecological sustainability.

3.3   INTENTIONS AND IMPROVEMENTS

As we've seen, FMEA and CA started out as a design tool for military weaponry. These tools can also be used as a process control tool, a prioritization matrix for corrective action, or even project planning. We will take a look at the historical intent of these uses, along with some examples of successful application. Our primary sources of comparison will be MIL-STD-1629 and QS 9000. These are the most prevalent sources for information on FMEA and CA, and yet they provide contrasting process methods. Learning from this historical foundation may ensure the most effective application of FMEA and CA toward social responsibility.

As a product design tool, both MIL-STD-1629 and QS 9000 propose that the FMEA and CA should be completed as early as possible during the product or system design phase. The earlier the analyses can be completed, the more effective, and less costly, remediation for identified risks will be. However, an iterative approach is necessary. For product design, the iteration between module or subsystem design, system functionality, risk assessment, feedback to subsystem design, corrective action, and modifications to system design is implied (Figure 3.4). Thus, the use of FMEA and CA as a product design tool requires a closed-loop feedback system and an iterative process of failure mode analysis. Early and often is the lesson for effectively using FMEA and CA as a method to develop designs robust against failure. This is a process, not a discrete, one-time action. The goal is a robust product design with an inherently low severity of failure, and reliable subsystems.

Historically, FMEA was used toward an additional application in QS 9000, as an analysis of potential process failure modes. The process FMEA is a way to communicate the recognition of process failure risk and the process control actions intended to mitigate this risk. The method in which the process FMEA is completed is identical to the design FMEA; however, the focus is on potential failures of the manufacturing process from intended function. These risks might include failure to deliver product on time, risks to the safety or health of personnel involved in manufacturing, or process failures that may result in defective product. The quantitative requirements of analysis of the occurrence of process failure modes required

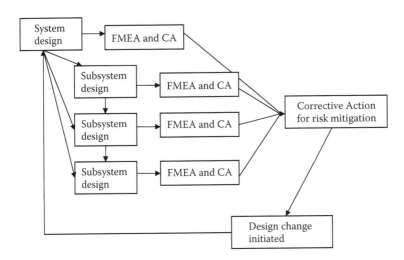

FIGURE 3.4 FMEA and CA as a part of the design process.

statistical analysis of the process capability, thus leading to additional information of the prediction and control of process parameters. The application of process FMEA, through the requirements of QS 9000, caused the further spread of FMEA, now beyond the reliability engineer's desk, and the design engineer's desk, to the manufacturing engineer and operations manager concerned with keeping the manufacturing processes up and running.

The goal of this book is to spread the use of FMEA and CA well beyond any engineer's desk. We also want to significantly simplify and demystify the process. Our contention is that anyone, anywhere, with or without an organizational program of social responsibility, can take action to improve his or her world. The FMEA and CA need not be complex. It is a very effective tool to identify a wide variety of potential failure modes in a process, the effects of those failures, and the probability of occurrence of the failures. When used toward social responsibility, this will help us focus continuous improvement efforts on the most important risks to responsible behavior.

Our desire is to bring back a focus on the whole process of risk assessment. We have personally witnessed ineffective FMEA work when the organization's focus is on merely having filled out a spreadsheet. We will go back to the narrative as goal statement, clarification of assumptions, ground rules, decisions, learning points, or operational definitions. The narrative statement as a vision of social responsibility is just as important as the spreadsheets of the FMEA and CA. The process of discussing the possible failures of responsible behavior, and working to mitigate the most risky failures, is much more effective than a lot of trivial action that misses the big issues. Teaching you how to not just talk about social responsibility but, either as an individual, a work group, or a whole organization, pick up and take action on the most imperative risks to social and ecological sustainability with a straightforward, easy-to-use tool is our goal. We want to greatly broaden the recognition of risk analysis as the first step in building awareness in order to take action toward continuous improvement, building a more socially responsible world.

3.4 HOW TO DO AN FMEA

3.4.1 BASIC ELEMENTS OF THE FMEA

Historically, the failure mode effects and analysis and then the subsequent critical-ity analysis comprise the analytic part of the larger risk assessment process. For the socially responsible failure mode effects and analysis (SRFMEA) these two steps become the FMEA and *responsibility* analysis (RA). Instead of assessing criticality of failures to the weapons system, we will be assessing the risks associated with irre-sponsible actions to society and the environment in the responsibility analysis. It is important to remember that the FMEA and RA are only two steps in the overall risk assessment process, which additionally includes the supplier-input-process-output-stakeholder (SIPOS) analysis, the narrative, and the continuous improvement plan (we'll explain this whole process in the next chapter). This section will focus on the analytic tools of the FMEA and the RA. The next chapter will go into further detail on other steps of the risk assessment.

3.4.2 THE ANALYSIS OF RISK

As we dive deeper into understanding the analytic tools of risk assessment, it is important to understand the intent, techniques, and methods of each step of analysis. Although most of the analysis in risk assessment is qualitative in nature, it is analysis nonetheless. It should be treated as carefully and precisely as any regression study or hypothesis test. But because of its qualitative nature, aspects such as operational definitions and triangulation of data become more important. As with any type of analysis, bad data going in results in bad data coming out. Having the right informa-tion and the right experts involved in the analysis is important.

Risk assessment is a snapshot in time. We can only analyze the risks present in the environment today and the system today. Tomorrow, both the environment and the system will be different. Caution should always be made when dealing with an "old" risk assessment. For this reason, your FMEAs and RAs should always prominently display a date. In fact, we recommend that your electronic files be named with the date on which the analysis took place. This provides clarity and a reminder to everyone using the analysis on the timeliness of the data involved. At any later date, the first question that should arise when using the analysis is: What has changed in the system and in the environment since the analysis took place?

Risk assessment can only be completed on risks that are known. And the knowl-edge of risk is always changing. This is another reason why timeliness, and under-standing FMEA and RA as a snapshot, is necessary. But beyond time concerns, the risks analyst needs to research what is currently known about the risks being assessed. For example, there was a time when it was not known that lead had a harmful effect on human childhood development. Manufacturers making, and build-ers using, lead paint were not acting irresponsibly. The risk was unknown. Only after significant social harm was done with the overabundant accumulation of lead in paint, toys, canned food, gasoline, and other fuels did scientists connect lead

Limitation	
Time	The risk assessment is based only on the current state of the process during the analysis. Tomorrow both the environment and the process will have changed. The time when the analysis is completed is a constraint on the risk assessment.
Knowledge	We can only assess those risks that are known today. There will always be unknown risks. Tomorrow new knowledge will uncover additional risks. The knowledge available at the time when the analysis is completed is a constraint on the risk assessment.

FIGURE 3.5 Risk assessment limitations.

poisoning to serious childhood developmental issues. Only after the risk was known were the manufacturers who used lead acting irresponsibly.

Sometimes it may take years, or even decades, for scientists to clarify the risk. For example, it is currently suspected that there is a pollutant responsible for some of the acceleration of cases of autism spectrum disorder. But, as of this writing, scientists have not yet determined what in our environment is responsible for the increase in incidents of this disease. Today, there is no ability to determine the risk of this unknown pollutant. However, as soon as the culprit chemical is known, organizations behaving with social responsiblity will be pulling out their risk assessments and revising their analyses in consideration of the new information.

The analysis of risk is constrained by time and knowledge (Figure 3.5). It is limited due to changes that will happen to the system and to the environment the minute after the analysis has occurred, and it is limited due to the ever-changing information on what comprises a risk. Providing clarity and attention to the date upon which the risk assessment was completed gives a warning to future users to investigate changes and new information before relying upon the information in the analysis.

3.4.3 PROACTIVE RISK ANALYSIS

Because of the inherent problems of old risk analysis data, risk assessment is best when used as a proactive tool. We have found that the best approach for risk assessment is from the opposite direction of risk, or reliability. If risk assessment is approached by analyzing where the system might be unreliable, as opposed to what are the risks to the system, a more proactively spirited analysis can be made. Risk is the probability of a loss or problem; unreliability is the lack of receiving some expected assurance (Figure 3.6). This is a subtle but important difference. Let us explain.

In recent years there have been reported cases of illness and death in the United States caused by vegetables tainted with potentially deadly bacteria. This is a serious consumer issue. For a farm, food preparation organization, grocery store, or restaurant, having tainted vegetables should be a risk documented in the FMEA and RA as an ISO 26000 core subject of "consumer issue." The risk side of the analysis provides the perspective of tainted food, with the desire of bacteria-free food. The reliability perspective would dictate the perspective of clean food with the desire of safe. These are two sides to the same coin. The approach of bacteria-free is from a

Risky	Reliable	Unreliable
Dangerous	Dependable	Defective
Perilous	Consistent	Unpredictable
Hazardous	Unfailing	Erratic
Precarious	Accurate	Inaccurate
Uncertain	Operational	Erroneous
Exposure	Functioning	Broken
Potential loss	Effective	Failed

FIGURE 3.6   Risky, reliable, and unreliable: the perspective of the risk analyst.

risk perspective, while the approach of safe is from a reliability perspective. Because the best risk analysis is done proactively, a reliability approach is preferred.

Now, this can be a little confusing, but bear with us. We always look at the failure modes, even when taking a reliability perspective. But the reliability perspective provides a more holistic improvement. Let's continue with the same example. If we take the risk perspective, our FMEA will call out the failure mode of *bacteria present* on the FMEA for the food in the field. There is nothing technically wrong with this perspective, but what about the reliability perspective? If we take the reliability perspective, we want to recognize the opposite of reliably safe food; we would show *unsafe* on our FMEA. There can be a lot of potential effects to *unsafe* beyond just bacteria laden. For example, one cause of *unsafe* might be if metal contamination broke away from the harvesting equipment and embedded in the broccoli. Metal shards in broccoli are certainly an unsafe consumer issue. However, if we only analyzed *bacteria*, we would not be cued to think of other safety issues beyond diseases. Taking a reliability perspective opens up the analysis to a more holistic recognition of causes and effects of failures.

Our discussion on concerns with timeliness and a reliability approach are broad in nature. In the next section we will get into more detail on the elements of the risk analysis: severity class, occurrence class, and risk priority number. Advice on accomplishing this analysis will be taken with the consideration that the analyst is conscious of the time-constrained validity of his or her analysis, and that the analyst will be taking a reliability approach to his or her work.

3.4.4 The Magnitude of Potential Losses: Severity Class

An important part of any risk assessment is the identification of potential losses. With FMEA, this magnitude of potential losses is referred to as severity class. In the original version of FMEA, severity is defined as the worst potential consequences of the particular failure mode under analysis. Our definition is the same, but with respect to social responsibility. We have to assume the system and environment as it exists today, during the analysis, and we have to assume that we are concerned with known risks. Upon these assumptions, severity is then evaluated based upon the loss of reliable socially responsible behavior. Remember, a reliability perspective gives us a more holistic evaluation of causes and effects. Our recommendation is that failure modes are segmented based on the seven ISO 26000 social responsibility core subjects: organizational governance, human rights, labor practices, environment, fair operating practices, consumer issues, and community involvement. For each of these aspects, failures in the system are identified, their harm to society documented, and then the severity of the impact of the harm quantified.

There are a couple of different ways to approach this quantification. The authors of the original MIL-STD-1629 used the intra-FMEA ranking technique. With this method, after all failure modes and their associated effects are identified, which usually results in many pages of spreadsheet, each effect is ranked within the single FMEA. The worst impact, or largest negative effect on society, is given the highest number. The least impactful failure mode is given the lowest number. The rankings are usually integers from 1 to 9, with 1 being the least harmful effect.

The spreadsheet, called "the FMEA" (see Appendix A), becomes a prioritization matrix. It will list all possible failures of socially responsible behavior, for each of the seven core subjects of ISO 26000. For each failure mode, there may be many effects of that failure listed. After all modes and effects are listed, each failure effect will be rated, from 1 to 9, for its relative severity, compared to all other failure effects in that FMEA, thus creating an intra-FMEA ranking of failure modes, applicable only to the process under analysis.

If FMEAs across the organization will need to be compared, then a process of rating with organization-wide operational definitions for severity class is recommended. In this method of quantifying severity, the organization first sets out to determine examples of the highest rating for severity class. For the FMEA applied to social responsibility, this is needed for each of the seven categories of social responsibility. A human rights example is shown in Figure 3.7. In this example, the worst possible impact is a catastrophic failure in that category. For example, a catastrophic impact on human rights might be death of dissenters, or life long imprisonment or slavery. The organization, given the society in which it operates, and the narrative of the process, needs to decide what the outcome resulting in "catastrophic" failure of human rights means. Then catastrophic failure for each of the six other categories also needs to be defined. This then creates organization-wide templates for each rating for each category. Having these definitions in advance facilitates the completion of the risk analysis as well as providing consistency, FMEA to FMEA and process to process.

The former ranking technique, using an intraprocess ranking of least to most risk, eliminates the inevitable arguments over ratings, but it prevents the comparison of one FMEA with another. This technique certainly creates a prioritization for improvement action. However, within an organization, the process for metal joining and the process for company travel cannot be compared. If each FMEA and RA will be acted upon within functional silos in the organization, this limitation may be acceptable. However, if the whole organization is interested in measuring its overall improvement of risk abatement, the intra-FMEA ranking will not be effective.

One caution in the creation of common, organization-wide definitions for severity class is to remember that the ultimate goal of the risk analysis is the prioritization of risk in order to take action. Assigning quantitative values to subjective assessments of severity class is only beneficial when discrimination between impacts is

Rating	Severity of Human Rights Risks
1	• Unlikely impact on civil, political, economic, social, work, or cultural rights • No risk of discrimination or complicity • Attention to resolving grievances and due diligence
2	• Unlikely impact on civil, political, economic, social, work, or cultural rights • No risk of discrimination or complicity • Some lapse of resolving grievances and due diligence
3	• Unlikely impact on civil, political, economic, social, work, or cultural rights • Minor risk of discrimination or complicity • Some lapse of resolving grievances and due diligence
4	• Unlikely impact on civil, political, economic, social, work, or cultural rights • Minor risk of discrimination or complicity • Little to no attention to resolving grievances and due diligence
5	• Unlikely impact on civil, political, economic, social, work, or cultural rights • Some risk of discrimination or complicity • Little to no attention to resolving grievances and due diligence
6	• Possible impact on civil, political, economic, social, work, or cultural rights • Some risk of discrimination or complicity • No attention to resolving grievances and due diligence
7	• Possible impact on civil, political, economic, social, work, or cultural rights • Possible discrimination or complicity • No attention to resolving grievances and due diligence
8	• Probable impact on civil, political, economic, social, work, or cultural rights • Probable discrimination or complicity • No attention to resolving grievances and due diligence
9	• Significant impact on civil, political, economic, social, work, or cultural rights • Probable discrimination or complicity • No attention to resolving grievances and due diligence

FIGURE 3.7   A severity rating table: human rights example.

made. If everything is given a rating of 9, or a rating of 5, or a rating of 1, then no prioritization is made. For this reason, we've seen the best definitions of severity class include observable, tangible conditions. Observably anchored rating scales are needed; ensure that your organization's definitions of severity ratings are observable potential losses that describe the magnitude of the loss.

3.4.5   THE PROBABILITY OF LOSS: OCCURRENCE

In the FMEA style of risk assessment the probability of loss is analyzed through the use of an occurrence class rating. The analytic quantification in the process of social responsibility risk assessment is the assigning of occurrence class to the responsibility analysis. It is important to remember that we are ultimately quantifying the prediction of failure; the failure rate or occurrence of failure can never be exactly known. Historically, MIL-STD-1629 recognized two different ways to assign occurrence ratings: qualitative and quantitative. In the quantitative analysis statistics on the process capability or product reliability were used in determining the predicted occurrence of failure. If it is known that the vacuum cleaner has a mean time to failure of 1,357 hours of use, based on laboratory reliability testing, then we should use this information in our FMEA to consider the probability of occurrence of failure modes. The second method is to use a qualitative assessment of the probability of occurrence. This method uses a simple five-category rating: frequent, reasonably probable, occasional, remote, and extremely unlikely. Each categorization of the probability of occurrence of a failure mode will have an occurrence rating assigned (Figure 3.8). When preparing a criticality analysis on a mechanical system for system failure, if there is laboratory or field mortality data, the quantitative method may be used. When preparing a responsibility analysis on an organizational system in need of improving socially responsible behavior, the qualitative method may be more useful.

When conducting a responsibility analysis on social responsibility aspects, we recommend the qualitative approach. Using the quantitative approach would require conducting significant research in the areas of the social impact. For example, if a failure mode identified human rights violations resulting in the effect of human smuggling, the analyst could research the occurrence rate of human smuggling in that specific situation. There may be border patrol data on the statistics of known occurrences of smuggling. There are occasions when hard data are available. If so, use them. However, we anticipate the availability, or even accuracy, of these types of statistics to be extremely rare. And anyway, the SRFMEA will be rating the occurrence of social impact from consumer issues, human rights, and the environment in the same analysis. Therefore, knowing the exact probability of human smuggling, translated into a quantitative rating, is no more accurate than applying a qualitative rating.

Like severity class, when preparing to compare RAs across the organization, having an organization-wide operational definition for occurrence class is helpful. The organization should decide: What is frequent? What is reasonably probable? *Frequent* may be very different from one organization to another, or from one context to another. A "frequent" occurrence of human smuggling in Mexico may carry a very different definition than the same rating of a different organization in Finland. Consumer issues may have a very different definition of *occurrence* for an industrial chemical distribution company versus a manufacturer of children's toys. In Figure 3.9, we have continued the human rights example (shown in Figure 3.7) to demonstrate how an organization might define organization-wide definitions to *occurrence*. These ratings, shown in Figure 3.9, may be shocking and dismal for

Rating	Probability of Occurrence
9	Frequent
7	Reasonably probable
5	Occasional
3	Remote
1	Extremely unlikely

FIGURE 3.8   Qualitative occurrence ratings.

Rating	Severity of Human Rights Risks
1	• Extremely rare human rights violations • No known practice of bribery or discrimination • Rarely, grievances go unaddressed
2	• Extremely rare human rights violations • Occasional practice (less than once per year) of bribery or discrimination • Occasionally, grievances go unaddressed
3	• Extremely rare human rights violations • Occasional practice (less than once per year) of bribery or discrimination • Frequently, grievances go unaddressed
4	• Extremely rare human rights violations • Monthly incidents of bribery only at the lowest levels of the organization • Rare incidents of social liberty violations
5	• At least one instance of civil or political discrimination annually • Monthly incidents of bribery only at the lowest levels of the organization • Rare incidents of social liberty violations
6	• Occasional instances of civil or political discrimination annually • Frequent incidents of bribery only at the lowest levels of the organization • Occasional incidents of social liberty violations
7	• Monthly incidents of bribery at multiple levels in the organization • Monthly occurrences of civil or political discrimination annually • Occasional incidents of social liberty violations
8	• Daily incidents of bribery at multiple levels in the organization • Monthly occurrence of civil and political discrimination • Occasional incidents of social liberty violations
9	• Daily incidents of bribery at multiple levels in the organization • Monthly occurrence of civil and political discrimination • Daily incidents of social liberty violations

FIGURE 3.9   Organization-wide occurrence rating table for human rights.

most first-world countries; however, the same ratings may be very reasonable for some third-world countries engaged in the battle for human rights. The benefit of any organization-wide rating is the ability to compare FMEAs for risk priorities across the organization. It is equally valid, and in some instances more valuable, to simply rank order the frequency of occurrence within an FMEA, giving the failure effect with the least probability of occurrence a rating of 1, and the most probable a rating of 9. However, with this method of rating, different FMEAs should never be compared to each other.

When completing an FMEA we're building a prioritization matrix. We can only work on improving a few important aspects and impacts at a time. We are using the rating of occurrence to prioritize working on the most frequently occurring impact first. The method your organization chooses should be the most effective and efficient toward this goal. If you will be applying mixed resources across the organization, use organization-wide standard definitions of severity class and occurrence class. If your improvement resources will be found within the departments and functions independently, then a simple ranking of most risk to least risk within a single FMEA is an acceptable rating technique.

3.4.6 RISK PRIORITY NUMBER

The compilation of the risk priority number (RPN) takes into consideration both the severity rating and the occurrence rating. Ultimately, it is the risk priority number that becomes the target of attention when selecting the most important risk for continuous improvement action. The risk priority number did not exist in the original MIL-STD-1629. This is an artifact of the AIAG changes to the original intent. In the AIAG version of FMEA, the purpose of the RPN is to provide the ultimate prioritization rating on the individual failure effect. This would result in hundreds of RPN scores for a typical process FMEA.

For the social responsibility failure mode effects and analysis (SRFMEA), we do recommend using an RPN, however, in a manner different from AIAG. The culmination of the responsibility analysis should be an overall priority ranking of social responsibility failure modes of only the highest severity class. The SRFMEA team should decide, when transitioning from the FMEA to the RA, how many failure modes to consider in the RA. We do not recommend moving all failure modes forward to the RA. Chapter 4 will discuss this step in more detail. For example, an RA may only have 10 to 20 failure modes and effects, out of 70 to 100 identified in the FMEA, to which occurrence class will be applied. In the RA the RPN is then compiled for only these failure modes. The RPN then provides an easy, quantitative method to prioritize failure modes for action.

We will discuss two different methods to use when calculating the RPN. At this point the severity class and the occurrence class columns are complete on the RA (see Appendix B). No RPN should be calculated until all failure effects are rated for severity. After the table is complete, then RPN can be assigned in one of two mathematical methods. The first is a simple multiplication of severity times occurrence. In this method, RPN is the product of the severity class and the occurrence class. The second method is to make a compound number from the two numbers. For example, if severity class is 9 and occurrence class is 4, then the RPN is 94. This method works best if the ratings for severity and occurrence are limited to integers between 1 and 9.

There are pros and cons to each of these methods for calculating RPN. In the product method, large numbers result, the maximum being 100 (10×10). In general, we find that there is larger discrimination between individual RPN results with this multiplication method. However, the number loses its connective discernment to its original severity and occurrence. In the second method, when the RPN is a compound number, the largest rating is 99, but each digit in the number has a meaning. Just by reading the RPN we are informed that the severity is 9 and the occurrence is 9. The compound number method of compiling RPN retains its meaning to the original severity and occurrence.

In most cases, we prefer the compound number method. With this method, an RPN of 92 can be easily interpreted that the severity is a 9, or maximum, and the occurrence is a 2. In this example, we have a failure mode that is very severe, but not very likely to occur. With the multiplication method the RPN, made from a failure mode with a severity of 9 and an occurrence of 2, has the same RPN as a failure mode with a severity of 2 and an occurrence of 9. A failure mode with a severity of 2 is not a very big impact, but with an occurrence of 9, this relatively benign irresponsible behavior can happen frequently. In most cases this is much less of a priority than a situation in which the impact is severe, although not frequently occurring. When preventing harm to society, we should concentrate on the highest severity first. In the previous example, using the compound number method would result in two different RPNs: 92 and 29. These are very different priorities! Clearly we should work on the risk of a 9 severity first. For social responsibility, the compound number method provides better differentiation between failure modes.

We want to provide one final caution when considering RPNs. In our experience, it is never effective to set cutoff scores. For example, we've seen organizations that say that there should be an improvement action plan for every RPN over 75. Arbitrary cutoffs like this present two negative results. First, for SRFMEAs for which there are only low RPNs no improvement is motivated. It is quite possible that an SRFMEA is done adequately and accurately, with every RPN falling below 75. With a cutoff criteria, the SRFMEA gets moved to the bookshelf with no action taken. Improvement opportunities are missed. A second, more destructive, thing can happen when cutoffs are given: The RPNs are "gamed" to always be lower than the cutoff. If the overwhelmed FMEA team is given a cutoff, there is a motivation to underestimate severity class and occurrence class to squeak below the minimum score required for action.

Both of these issues cause the organization to miss the whole purpose of doing the SRFMEA in the first place, which is a rational, reasonable, but conscientious improvement of social responsibility. We don't want too many improvement action items—that will only overwhelm the organization, and we don't want too few improvement action items—that will not prompt continuous improvement of socially responsible performance. We want a balance. Don't present cutoff criteria for your organization's SRFMEAs. Allow the FMEA team to choose which failure modes need attention, and how many action items they are capable of improving.

3.4.7 THE LEVERS OF RISK ABATEMENT

As the risk analysis is developed, it becomes apparent that there are some key levers for lowering RPN. As just discussed, the primary lever is severity class. Any action that makes a process input, output, or waste less severe to society should be a priority. While evaluating the improvement actions for a high-severity class failure mode, eliminating the failure mode takes top priority. For an example of improvement actions, see Chapter 6. The second lever of improving social responsibility is occurrence class. Actions that can be taken to reduce the frequency of impact or lessen the probability of occurrence are effective measures. So, lessen the severity and reduce the occurrence are obvious levers in the improvement of social responsibility.

There is a third lever lurking in the shadows—awareness. In the AIAG form of FMEA, a third factor is considered in the calculation of RPN: detection class. This is defined as the ability to detect a failure, and ideally take remedial action, before the failure has a negative effect. An example is the engine warning light on the dashboard of your automobile. Even if engine overheating failures cannot be lessened in severity, or reduced in occurrence, the ability to detect the engine malfunction and seek immediate resolution will prevent the effect of a cracked engine block. The original MIL-STD-1629 did not consider detection class as a valid factor in a risk assessment; it did, however, advocate a "maintainability assessment," which would address the engine warning light example. The original authors of the military standard were more focused on eliminating the cause of the failure, early in the design cycle. Our advice is to follow the latter approach with SRFMEA. We have seen too many poorly performed risk improvement projects that did nothing to enhance the reliability of the system, but rather focused on adding all sorts of detection bells and whistles to improve the RPN. This misses the intent of building reliability into the system.

Before we get too zealous on ignoring the detection class aspect of FMEA, there is an important kernel of knowledge to be gained from this aspect. Sometimes, the biggest improvement to social responsibility gained through the completion of an SRFMEA is an awareness of the system, the environment, and the risks. Improved awareness, in a way, is like improved detection class. So the third lever in the SRFMEA risk abatement is awareness—awareness gained from the act of performing the risk analysis. If the SRFMEA does nothing more than improve awareness of social responsibility factors and associated risks, it will have improved the organization's performance.

3.4.8   THE TWO-STEP PROCESS OF RISK ANALYSIS

In Chapter 4, we'll go into detail to describe what is required, the function, the reason, and the process in each step of the social responsibility risk assessment. Before we get there, let us say a few words about the importance of the two-step process of the risk analysis. Step 1 is the FMEA, and step 2 is the responsibility analysis (RA). Like many of the methods we're advocating in the SRFMEA, this two-step process finds its roots in MIL-STD-1629, and is contrary to the AIAG method of FMEA. In the original military standard the failure mode and effects analysis is completed first. This step culminates in the severity class or severity rating. Then, in a second step, the severity class is carried over to the criticality analysis. This is where the occurrence class is given and the RPN is compiled. This RPN is a resultant of the ranking of the severity and occurrence, and the source of finding the priority failure modes and effects for corrective action.

In the AIAG model, the severity, occurrence, and detection are all scored, and RPNs calculated, in a single step. The AIAG method of FMEA does not recognize a multistep process. We prefer the two-step process for several reasons. First, trying to do everything all at once is confusing. We've seen 70- and 80-page AIAG-style FMEAs. Now, try to keep track of all of those factors and resultants. Doing all ratings and RPNs in a single step seems to end up in a huge jumble of effects, some significant, some not, with RPNs being calculated as lines in the spreadsheet are completed. In our opinion, this isn't effective prioritization; rather, it is a task of "let's fill out the spreadsheet." If the failure modes and effects are first documented, it will become apparent that some risks identified are simply insignificant, redundant, or outside of the scope of the narrative. In a two-step process, these insignificant failure effects do not need to be translated to the responsibility analysis. Now, we're not advocating only transfer of the highest-severity failure effects; you may end up with a 49 and a 52 (using the compound number method), and you may decide to improve the failure mode with an RPN of 49, before you work on the one with an RPN of 52. But, there is no need to transfer failure effects with a severity class of 1, or 2, or maybe even 3, when there are dozens at 7, 8, and 9. This results in a more streamlined responsibility analysis. It's easier to stay focused on the prioritization for improvement, rather than filling out every line on the spreadsheet.

Another benefit of the two-step process is the improved ability to focus on that part of the risk analysis at the moment. The failure mode effects and analysis is all about brainstorming an exhaustive list of potential risk, thus improving awareness of all known social responsibility risks, and then ranking each by order of severity. That's all. There's no need to get overwhelmed with "what are we going to do" or "we have too much to do." The team can stay focused on identifying all the risks. After the list is complete, set it aside for a few days, then come back to do the responsibility analysis. Now focus on just that aspect of the risk analysis. There's no need to brainstorm more risks, now the focus is on prioritization. Here, we're using frequency of occurrence, combined with the already determined severity, to make a prioritization matrix, and only on those failure modes that were important to the narrative—a smaller list than found in the FMEA. The responsibility analysis is the "reality check." The FMEA sends you to the brink with "oh my gosh look at all the risks," and the responsibility analysis brings you back with "well first we need to work on these few priorities." The two-step process facilitates this focus.

3.5 WHY FMEA FOR SOCIAL RESPONSIBILITY

There are many benefits of using an FMEA format for capturing, ranking, and tracking potential failure modes for social responsibility. The irony is not lost on us—that a tool originally designed to ensure the deadliness of weapons systems is now being advocated for social and ecological sustainability. The benefits of capturing all of the potential failure modes, effects, and causes result in an extensive knowledge base of the systems, products, and components within a project, product, or process. The cascading aspect of communicating this information to all stakeholders within the supply chain ensures that all aspects of potential failures are identified at each step in the process. This information can begin the foundation of transparent communication with stakeholders. This knowledge base can be used as a training tool for new employees and be utilized to historically document new failures and changes to the product and processes over time. As new processes, products, services, or systems are added to the organization, this tool allows for the capture of the failures for these new elements.

A benefit of ranking failure modes is its use as a tool for identifying which failures have the greatest impact. Based on the ranking of highest priority, a listing of the issues allows the team to prioritize the risks that are most impactful toward social responsibility performance and begin to address and lower the scores with corrective, continuous improvement actions. This ranking also suggests the type and degree of resources that will be required to mitigate the risks associated with the various failures, and when completed early enough in the design or process deployment, negative impacts may be avoided.

The tracking nature of the FMEA allows for the team to follow up on those high-risk effects and to document the responsible party for improvements to the performance. The ability to rank and classify the failure modes that impact the social responsibility core subjects of the process allows the team to identify which of the core subjects is most important for their own organization. The FMEA format also allows for a way to document the risk mitigation measures and to measure continuous improvement success. As actions take place to improve products or processes by reducing severity or occurrence, the FMEA and RA, as living documents, should be updated to reflect these improvements. Over time the organization will see reductions in its RPNs.

A cross-functional team consisting of the various groups involved in the process (engineering, purchasing, health and safety, human resources, new product development, etc.) is encouraged to complete the FMEA form. The advantage of utilizing a cross-functional team approach is the varied experience and perspectives that each individual brings to the table. Although increased team diversity often leads to exciting and intense discussions when rating severity and occurrence, this difficulty in building consensus creates thoughtful debate about the organization's role in improving socially responsible behavior.

While the goal ultimately is to build robust organizational systems that ensure socially responsible behavior, the SRFMEA process can be utilized by a variety of stakeholders, throughout the organization, and for a variety of purposes. If there is a new process that involves a new service being offered, the SRFMEA process offers a proactive approach to evaluating potential failures before they occur. If there is a change to an existing process, the change can be incorporated and evaluated using the SRFMEA process to capture the change and its potential failure modes. If there is a current product that will be applied within the market in a new way, this also can be evaluated using the SRFMEA process. So the SRFMEA can be used to plan robust new processes, improve existing processes, or incorporate changes to new uses of existing processes. Utilization of effective risk assessment can improve a firm's image and competitiveness, because of the knowledge that the firm has on the unique nature of its product designs and processes. Utilization of the SRFMEA can serve as a means for improving customer satisfaction. A customer's knowledge that its supplier understands all aspects of the possible failures related to its product, and that a continuous improvement process is in place, gives the customer the assurance that risks are being mitigated, and the use of the SRFMEA format early in the design stage for the product or process can aid in the reduction of costs. The costs associated with finding out late into the development, launch, or production processes that there are potential failure modes that have not been addressed can be catastrophic. As the competency in the organization increases toward the use of SRFMEA as risk assessment, not only will social responsibility behavior improve, but also risk mitigation across other aspects of the organization may improve.

A key benefit of using the SRFMEA process is the continuous improvement nature of the tool. The SRFMEA is meant to be a series of living documents that change with any change in product, process, new failure mode, new technique, or new environment. The SRFMEA, if actively utilized in an ongoing manner, can be a timely record of knowledge, a foundation of a knowledge management system, and the source of validation of continuous improvement. The process allows for the tracking of action items that lower the risk of failure. It can become a source of pride for the team to witness lower and lower RPNs as the continuous improvement effort progresses.

3.6   FMEA CHALLENGES

There are many challenges to the effective use of SRFMEA. Since risk assessment is such a well-worn tool, we have a lot of experience, both good and bad, upon which to understand its effectiveness. Some of the cautions include an incorrect focus on the spreadsheet instead of the teamwork and continuous improvement actions; rather than being a process, it becomes a dusty document on a bookshelf—a "check mark" instead of a genuine investigative tool. If your organization is going to invest in the time and effort to tackle an ambition such as a social responsibility performance improvement initiative, use the SRFMEA wisely. Wasting the effort given to the process is, in itself, irresponsible!

The completion of the SRFMEA format is no guarantee that actions or improvements occur as a result. One of the key criticisms of the SRFMEA format is that there is no element of the process that ensures actions are implemented and reviewed for completion. If there is no dedicated resource with the responsibility of being the champion of change stemming from the SRFMEA, then there is a much lower chance of success in driving the improvements. Just as with any continuous improvement effort, there is no guarantee that just because the analysis is completed, the high-risk elements are actually addressed.

If the appropriate team members are not involved in the SRFMEA process, there is a greater probability that important failure modes and their effects will not have been identified. Important risks will not be known, documented, or acted upon. The entire intent of exercising the SRFMEA process is for the analysis to take place as early as possible in the design of processes and products. The identification of risks of failure late in the deployment may lead to costly redesigns and corrections, and having people involved in the SRFMEA activity who are not experts in the risks inherent in the product or process will not facilitate the mitigation of risk.

The best situation for completing an SRFMEA analysis is to have a cross-functional team ensure that all products, processes, and key areas of risks are addressed. It is disastrous for one person to complete the entire SRFMEA in a vacuum. The SRFMEA process is a team-based method. The interaction of the team members, gaining knowledge and awareness of risks, debating the probability of occurrences, and then being motivated to collaborate on solutions to mitigate the risks, is a benefit to the organizational culture. The interactions inherent in the SRFMEA process will build open, trusting, learning-oriented organizations.

The primary challenge to completing an SRFMEA is to avoid approaching the task as just the completion of a document, but rather to use the whole process, in the context of a cross-functional team, as early in the design phase as possible, to

When to use:

- You are developing a product, process, service, or project and want to understand the impact on society, the environment, and economics
- You need a way to capture and identify all potential elements of risk with a new product, process, service, or project
- You need a way to prioritize what aspects of a product, process, service, or project are high risk
- You need a tool that will allow you to capture, track, and monitor the status of actions related to failure modes and risk analysis
- You need to improve the reliability or quality of a particular product
- You need a process of performance analysis that is more proactive than reactive

When not to use:

- You are looking to analyze something other than potential failures of a product, process, service, or project
- You do not have a structure to support resources participating in the cross-functional nature of completing the FMEA
- You do not have a structure to support continual monitoring and updates to actions identified for improvements
- The target process for improvement is undergoing change

FIGURE 3.10 When and when not to use SRFMEA.

mitigate the biggest risks to performance. There is a time and place for SRFMEA; sometimes, we may not want to conduct this activity (Figure 3.10). If the organization is unable to honestly face risk, if the culture is such that problems are to be hidden, then this is not an effective climate for SRFMEA. If the organization is experiencing significant process change, or significant leadership shifts, it is not the right time for SRFMEA. If attention and allocation of resources will be distracted, or if resources are consumed with "fire fighting" just to keep the operation alive, it is not the right place for SRFMEA. Resources will not be available for the improvement actions identified. There is a need for organizational readiness to face the risks and engage in the continuous improvement. Start your SRFMEA journey at the right time and place, and with the right organizational intent.

4 The SRFMEA Process

What to expect:

- SRFMEA seven-step process
  - SIPOS
  - Narrative
  - Value/failure diagram
  - Value/failure focus
  - FMEA for social responsibility
  - Responsibility analysis
  - Action planning and tracking

4.1 THE "HOW TO"

This chapter outlines the task items and sequence for the completion of a social responsibility risk assessment through a social responsibility failure mode effects and analysis (SRFMEA). It is important to follow the process completely. We recognize that many people will be tempted to skip straight to the completion of the spreadsheets in this process. This is a big mistake. It is critically important to complete the organization's narrative to set up goals and focus for the rest of the process. It is also important to complete the suppliers-inputs-process-outputs-stakeholders (SIPOS) to identify the stakeholders of focus. Understanding stakeholder concerns is a critical part of social responsibility; skipping this step, or assuming that you know what stakeholders want, is a fatal error. Please dedicate your team to completing all the steps in the process.

This chapter will discuss each step of the process, the reasons for that step, the key tasks to deliver in that step, and cautions for the step. If your team gets stuck during the process of completing an SRFMEA, come back to this chapter and examine the overall process shown in Figure 4.1. After you've completed each task, come back to this chapter to ensure that you've addressed all of the cautions for the step. We've also provided a handy set of checklists in Appendices D to F. We want to make sure that each step is completed thoroughly before moving on to the next step. Moving forward haphazardly or too quickly will only cause problems in subsequent steps. Complete your risk assessment carefully and mindfully.

Most of the SRFMEA will be completed collectively by a team. Consider carefully who will comprise this team. Clearly experts on the process of evaluation need to be involved. Experts in the aspects of social responsibility will be helpful. Stakeholder representatives can also be important members of the team. Another team member, usually overlooked, is the "fresh eyes" person. Choose someone unfamiliar with the process, social responsibility, or the stakeholder to provide objectivity to the process.

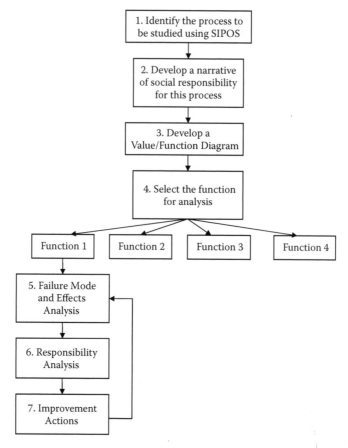

FIGURE 4.1   Generic process flow.

Consider having a formal facilitator assigned to the team. This person is not a part of the team, but he or she is present during all of the team's collective work. The facilitator's responsibility is to keep the team's process moving, such that conflict avoidance doesn't happen, and so that decisions are made efficiently and effectively. The diversity of thought process brought about through teamwork is a valuable addition to the SRFMEA process. If you find yourself completing an SRFMEA alone, at your desk, you're not doing it right!

In this chapter we will show you how to complete the whole risk assessment. In Chapter 6 we will show you a case study of an organization completing an SRFMEA. Also, refer to Appendices H to J for additional examples of the SRFMEA process. We will demonstrate the outcomes for each step of the process. You may find yourself flipping back and forth between these two sections as you move through the book, and that's OK. Each section of the following instructions includes (1) the reason for the step in consideration of the "big picture" of the risk assessment; (2) the deliverable tasks for that step, often including a template or form for completion; and (3) a checklist of cautions, as things to avoid before moving on to the next section.

Again, avoid the temptation to skip a step. The process flow is intentional, with each step providing important work to prepare information for the next step.

4.2   STEP 1: IDENTIFY THE PROCESS TO BE STUDIED USING SIPOS

4.2.1   REASON

The first step in the risk assessment is to identify the process of study in order to focus the work. This is a powerful step in the process. In our experience, many organizations do not accomplish great improvement in social responsibility because they don't know where to focus. Or, we'll find organizations that may do a great job on consumer issues, but ignore human rights issues. Start the SRFMEA by determining which organizational process will be studied. Taking a process-oriented approach allows for discrete focus, thus avoiding the overwhelming concerns of trying to improve the whole organization at once. Also, taking a process-oriented approach allows for the holistic analysis of risk on all aspects of social responsibility for that process. Because the team is focused on only one process, it isn't overwhelming to consider all aspects of social responsibility.

There is a con to taking a process-oriented approach. Care must be taken that cross-process comparisons are made. Performing risk assessment on processes in isolation will not ensure that the whole organization achieves improvements in social responsibility. For example, it makes no sense for an organization to complete a risk assessment thoroughly and completely in the manufacturing organization and ignore the office processes. Reducing fuel waste in the heat-treat furnace operation, while wasting enormous amounts of fuel on unneeded executive travel, does not represent a balanced, socially responsible organization. It is our suggestion that the organization's continuous improvement or corporate affairs leaders ensure that methods are deployed to balance the risk assessment work across processes throughout the enterprise.

Once the process of study has been selected and the team assembled, the first task for the team is the completion of the suppliers-inputs-process-outputs-stakeholders (SIPOS) analysis, shown in Figure 4.2. This is a tool taken from Six Sigma methods. Its intent is to provide, through a graphical depiction, a high-level understanding of the process of study. This gives the whole team a common understanding of the boundaries of the process of study. We will also use the stakeholder list created in the SIPOS in the next step. Don't worry about getting into too much detail in the process part of the SIPOS; this will also be covered in the next step. The primary reason for the SIPOS is the common understanding of the study focus and the identification of the stakeholders.

In summary, taking a careful approach to identifying the process to study, the process to target for improvement, helps create a balance between the workload and the need to improve. Many organizations could be easily overwhelmed by all the processes and systems that need social responsibility performance improvement. And because the work may be so overwhelming, no action is taken. It is difficult to see where to start. Setting the boundaries of study to a single process helps the organization to get off the starting block with a reasonable amount of work. The nature

Step 1. Identify the process to be studied using SIPOS

Suppliers	Inputs	Process	Outputs	Stakeholders
Utilities service	Requests for reports	See Step 3.	Printed: reports, forms, graphs for use	Office personnel
Office supplies vendor	Electricity			Recipients of printed reports
	Machinery: copiers, computers		Used printouts to be recycled	
Machinery manufacturer	Report information format		Electronic media: files, reports, data, analysis	Community waste repository
Information systems	Communication devices: phones, computers		Batteries	

FIGURE 4.2   Example SIPOS: Office waste stream.

of continuous improvement is small, with steady improvements happening over and over again. Identifying the process to be studied using SIPOS helps the organization to see the path of continuous improvement.

4.2.2 THE TASK

4.2.2.1 A Checklist for SRFMEA Readiness

The primary task in Step 1 is ensuring that the organization is committed to the goal of social responsibility performance improvement, resources are ready to start the risk assessment, and the best process has been chosen for assessment. This includes making sure the right team members are selected, and that these team members have time allocated to devote to the upcoming work. This is also the time to consider the scale of the process chosen, the priority of this process in the grand scheme of all risk assessments, and the strategic implications of choosing this process over others. Considering the strategy of the organization and the priority of the process to consider will help ensure that leadership support for the risk assessment exists. We will need leadership support for the allocation of the resources, the interaction with the stakeholders, and most importantly, the investment required for the improvement actions. Complete the checklist in Figure 4.3. Every question should be answered in the positive.

4.2.2.2 The SIPOS

The SIPOS (pronounced sy-poss) should always be completed collectively by the whole team. Since the primary reason for the completion of the SIPOS is a common understanding, everyone should be involved. It is somewhat of a brainstorming effort and should be worked backwards. Using the format in Figure 4.2, the team should first brainstorm all of the stakeholders of the process under consideration. Consider organization stakeholders, employees, their families, vendors, customers, and stockholders. Consider community stakeholders, utilities, neighbors, schools, governments, and local businesses. After a complete list of stakeholders has been created, the team should brainstorm all the process outputs that impact these stakeholders. These may

Has an organizational strategy been developed toward social responsibility?	
Has a priority of processes to study been developed?	
Is this the most urgent process to study?	
Are there any significant changes planned for this process in the near future?	
Can expert resources be assigned to the risk assessment team at this time?	
Can expert resources be assigned to improvement actions at this time?	
Is senior leadership aware of this risk assessment? Is there agreement to proceed?	

FIGURE 4.3 A checklist for process selection.

include productive and wasteful outputs. Think about all of the material and energy that is an outcome of the process. Next, create a very brief, high-level process flow; this is only three or four process steps at most. The important part of this creating the process flow diagram is to identify the start and stop boundaries of the process of study. The interior detail process analysis will happen in the next step of the FMEA.

After the process has been sketched, the team should brainstorm the inputs necessary to the process. Think about all the materials, energy, information, and human resources required to conduct the process. Look for hidden inputs. You will list productive and wasteful inputs. Lastly, identify all of the suppliers of these inputs. There should be a supplier identified for every material, energy source, information, and human resource. These suppliers may become important resources during the creative identification of your improvement actions. Keep the SIPOS ready to view during the risk assessment. Stakeholders will be the recipients of improved performance, outputs will be targets for improvement and effects of failure modes, the process will be the subject of improvement, the inputs will be altered to make the improvements, and the suppliers will be involved in developing those improvements. Each part of the SIPOS will be important in subsequent steps.

4.2.3 CAUTIONS

The following list describes some cautions for this first step of the SRFMEA process, identifying the process to be studied. Avoiding these circumstances will help the SRFMEA effort.

- Failure to involve stakeholders in the SIPOS development
- Choosing a process that is unimportant to the organization's social responsibility strategy
- Choosing a process that is about to go through significant change
- Not allocating resources to the SRFMEA team

4.3 STEP 2: DEVELOP A NARRATIVE

4.3.1 REASON

The narrative is the story of the process becoming socially responsible. Now, for the technical, reliability engineer or risk analyst this may be an uncomfortable task. Its value will become obvious when the risk analysis is complete. At this time the organization has selected a process of study. This process has been chosen for its opportunity and its readiness for improvement considering the overall social responsibility strategy of the whole organization. And, at this time, the team has completed the SIPOS. Everyone on the team has the same basic understanding of the process, and all the stakeholders have been identified. Now, the narrative answers the question "So what?" A description of what the organization hopes to gain from improving this process is the goal of the narrative. Social responsibility covers a huge body of topics: Everything from environmental impact to labor relations will be considered. For many processes, more focused goals and objectives can be made for improvements. The narrative will present that focus.

Like so much of our advice in this book, the narrative takes its roots from MIL-STD-1629. This original form of FMEA required the team to set ground rules and assumptions before the risk analysis commenced. And this report of ground rules and assumptions was always considered part of the risk assessment, not just the spreadsheets of RPN numbers. Stopping to develop a narrative disciplines the team to record their thoughts, assumptions, biases, goals, aspirations, constraints, and strategies. It helps to create a body of knowledge of what the team was thinking as this "snapshot" of risk assessment was taken.

The narrative can take many forms. Most organizations with the dedication to complete an SRFMEA already have some sort of vision and mission statements. How the process of study contributes to the whole organization with respect to the vision and mission of social responsibility should be stated in the narrative. If any aspects of social responsibility are off limits, this should be stated in the narrative. All operational definitions of terms should be stated in the narrative. How the stakeholders were chosen, and if there are particular stakeholders of interest, should be documented. Putting on paper the thoughts of the risk analyzers is the goal of the narrative.

The narrative serves two purposes. It should have enough information to provide focus for the team. For example, at the completion of the narrative, the team may decide that neighbors are the stakeholders of interest, and environmental impact is the aspect of interest. From here the risk assessment will minimize the analytic considerations of customers as stakeholders and consumer issues as an aspect targeted for improvement. Besides providing focus, the narrative should also inspire. It should remind the team why their risk assessment is important to the improvement of social responsibility. It should provide information on the link between this work and the larger goals of the organization. It should move the team to mindful concern for the work ahead.

4.3.2   THE TASK

We've found the best way to complete a narrative is through a team interview process. This may include the team members interviewing each other, interviews with leaders in the organization, and interviews with the stakeholders. We've offered a series of potential interview questions in Figure 4.4. You may add to or change these questions. The interview questions should be tailored to both the organization and the process of study. After the questions are answered, use your creativity to develop

Step 2: Develop a narrative for social responsibility. In paragraph form, as a team, answer the following questions while creating a narrative of thoughts, goals, and patterns that are pertinent to the impending analysis.

Profile	Who are we as an organization? Who is on the team?
	What is and isn't included in this responsibility analysis?
	What is the planned timing of the analysis?
	When will improvements be planned for deployment?
	How will this analysis be transparent to the organization?
	What leadership commitments are being made toward this analysis?
Goal	What is the goal of this analysis?
	What is the vision statement for this process's future state?
Context	Why do we want to improve social responsibility?
	Which of the seven aspects will be the focus? Will any aspect be intentionally eliminated from the analysis?
	Is this a part of a bigger program? If so, how?
	How will changes from this analysis change the social responsibility strategy of the organization?
	How will this analysis fit into the larger organization?
Stakeholders	Who are the stakeholders?
	How will you talk to them?
	How will you monitor and measure changes of outcomes to them?
	Are there stakeholders whose interests will be minimized during the risk assessment?
Risks	What outcomes do the stakeholders receive? And what is the potential impact to them—positive and negative?
	What will be considered the qualitative levels of severity of failure mode? How bad is bad?
	What will be considered the qualitative levels of severity of probability of occurrence? How bad is bad?
	How will awareness of risk be found and communicated?
	How will best practice be found and communicated?
Process of analysis	What are the functions of the process to be analyzed?
	What are the material input, energy input, human input, and waste input?
	What are the outcomes of the process to be analyzed?
	What are the material outcomes, energy outcomes, human outcomes, and waste outcomes?
	What are the boundaries of the process of analysis? Where in the process flow does the analysis begin? End?

FIGURE 4.4 Questions to answer in the narrative.

Summary of analysis	(To be added to the narrative after the analysis is complete)
	What priorities of improvement were found?
	What improvement actions were found to be needed?
	How will progress of improvement be measured? By whom?
	How will organizational governance ensure improvement actions happen?
	How will the findings of this analysis become a part of the organization's knowledge management system?
	Who is the process owner?
	At what frequency, and with whose leadership, will regular, frequent reviews of this analysis continue?
	What unusual conditions should prompt reassessment?
	How will the findings of this analysis become feedback to potential strategy change of the organization?

FIGURE 4.4   (Continued)

the narrative. Some organizations may create a story using information from the interview. We've seen others create graphical depictions or murals that exemplify the narrative. Others may simply document the narrative in a question-answer table. Choose the appropriate method for your organization.

We start the narrative with a profile statement. This tells the story of how this risk analysis fits into the bigger picture. We talk about the team, leadership commitment, and why the organization is investing in this risk assessment. Next is a statement about the goals. Here the hopes and aspirations upon completion of this risk abatement activity are added. Expressing the expected benefits of the improvement activities, although yet undetermined, could be a part of this section. The context section of the narrative speaks to the assumptions and situation in which the SFRMEA is occurring. This should be where focus is provided, exclusions stated, and definitions created. When writing on the context of the situation, it is important to remember that any changes to this process will have an impact on adjacent processes. Presenting possible constraints due to this potential chain reaction could be stated in this section. This section may be the riskiest part of the narrative. Let's face it: We live in a litigious world. Your risk assessment may uncover risks heretofore unknown. That's a risk in even completing the risk assessment. The risk section of the narrative should address this. How will you be working with your stakeholders to communicate and provide assurances based upon your findings? The narrative should include descriptions of the stakeholders, reasons for choosing those stakeholders, and their interest in the process of study. The narrative is also the place to document any quantitative rating scales. How bad is bad? The team needs to think about this before they dive into the risk analysis. Next, clarity on the process of analysis is presented in the narrative. This has been done pictorially in the SIPOS. A brief statement of where the process of study starts and stops is stated.

After the SRFMEA is complete you will come back to the narrative to tell the rest of the story. In the spirit of continuous improvement, we should conceive of every risk assessment as one step in a series of improving risk assessments. The summary of analysis section is provided to document the outcomes determined, in preparation for the next team to take the next steps of improvement on this process. Coming back to the narrative to do this step also provides structure for testing the improvements against the original goals. Did you achieve them? This task may expose a continued gap.

However you decide to creatively document your narrative, it will be a critical coalescing of information to prepare for the SRFMEA. The goals, context, operational definitions, stakeholder notations, and process clarifications will provide important information for the risk analysis. Coming back to the narrative after the risk analysis will provide testing of achievement. And, if goals are achieved, then the summary of analysis may be a great opportunity for celebration!

4.3.3   CAUTIONS

The following list describes some cautions for this second step of the SRFMEA process, developing the narrative. Avoiding these issues will ensure the team has focus, goals, and aspirations concerning the SRFMEA work, and it will ensure that the team's work is aligned with the organization's strategic initiatives.

- Failing to provide a focus for the team
- Failing to prepare to mitigate newly identified risks
- Failing to provide a connection to the organization's big picture for social responsibility
- Missing the opportunity to document the team's thought process on assumptions, definitions, goals, and constraints
- Not circling back to complete the summary of analysis

4.4   STEP 3: DEVELOP A VALUE/FAILURE DIAGRAM

4.4.1   REASON

The reason for the value/failure diagram (VFD) is to ensure that the whole SRFMEA team understands the process to be studied. However, the technique we suggest for completing the value/failure diagram may be a little different than any you've seen before. We have an example in Figure 4.5. The VFD is a cross between a value stream map and a functional flow diagram. Having chosen a process of sufficiently narrow focus during your SIPOS completion, we now state all of the steps of the selected process that create a *value function*.

Let's explain value function. Take your whole process. For almost all processes, during discrete phases of the operation of the process, value is created. For a manufacturing process this may be after each new component is assembled. For an office process this may be after each necessary task is delivered to the next internal customer. Look for process step outcomes; this may be a material product or a hand-off during the process. These are the value functions. They can be thought of as functions, in the process, that add value. We've given several examples in the next

Step 3. Develop a Value/Failure Diagram. Identify the failure modes for each value function in the process flow. Select only one function at a time for further Responsibility Analysis.

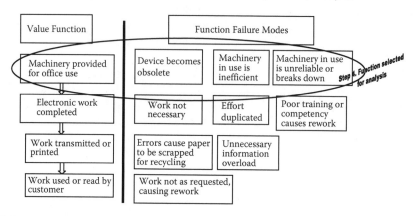

FIGURE 4.5   The value/failure diagram.

chapter. In order to assess the risk of all social responsibility failures, we will study these value functions as discrete elements.

For each value function, brainstorming will be completed on probable failure modes in that function. These are opportunities for identifying wasted material, wasted energy, and wasted human resources. Different value functions will have different potential failures. Identifying these may begin to sort very risky value functions from very benign value functions. Listing possible failure modes next to the value function creates a sort of voting, or frequency, which highlights those with more potential risk. We'll leave the benign value functions out of the risk analysis, thus providing additional focus for the risk analysis.

So the reason for the value/failure diagram is to understand the process in detail, understand the discrete value functions within the process, and lastly, to identify the most problematic value functions to focus subsequent efforts. Completing the VFD as a team ensures that everyone on the team has the same comprehension of the steps in the process under study. The VFD can also facilitate the prevention of "scope creep," by reminding the team where the focus of study begins and ends.

4.4.2   Task

Using the start and stop of process consideration from the SIPOS, identify the value functions within the process. Look for hand-offs of work from one person to another. Look for products delivered to internal customers. Look for the material change during manufacturing or assembly. These are indicators of discrete value functions. You are identifying all of the points of the addition of value to the product as it progresses through the process of study. Break the process flow in steps as small as possible. For example, in an assembly process, station 110 assembles three bolts and then passes the product to station 120, where a label is applied. Station 110 and station 120 are two different value functions. In an office example, if the financial report is pulled from the enterprise system by Mary, and then delivered to Sharon for controller approval before being delivered to the board of directors, then Mary's report assembly is one value function, and Sharon's approval is a second value function. You don't need to go so far as to describe the assembly steps of each individual bolt, or Mary's steps to extract the financial report from the system. Looking at discrete value functions will provide just the right amount of analytic resolution to detect the impacts to social responsibility.

Now, for each value function the team should brainstorm failure modes. Don't get too hung up on covering all seven core subjects of social responsibility. This is just high-level brainstorming. Think about wasted material, wasted energy, or wasted human resources. You might find a lot of duplication from one value function to the next in identified failure modes; that's OK. Think about where "things" go and "who" does what. What kind of energy is required? Wasted? These are potentials for failures.

After each value function has been reviewed for potential failure modes, the process flow diagram graphically depicts a kind of voting outcome. The most potentially problematic value functions will have the most failure modes identified. Use this information to refocus on just a few value functions, the most problematic value functions.

This style of value/function diagram presents the right amount of detail for the FMEA and RA, not too much, not too little. In many instances this may provide less granularity than a typical process flow diagram. And the identification of function failure modes, for each value function, will accelerate the next step of determining the focus of the FMEA and RA, which will allow the team to achieve continuous improvement of social responsibility.

4.4.3 CAUTIONS

The following list describes some cautions for this third step of the SRFMEA process, completing the value/function diagram. Avoiding these issues will ensure the team understands the process under study and begins to identify the most risky parts of the process.

- Not breaking the value functions down enough
- Not involving process users in the identification of the value functions
- Missing whole value functions
- Going beyond the boundaries of study identified in the SIPOS

4.5 STEP 4: FUNCTION SELECTED FOR ANALYSIS

4.5.1 REASON

The next step involves choosing the value function(s) that will be the focus of the FMEA and RA. We want to choose the steps in the process that present the most risk, and therefore are the most important targets for improvement. But, at the same time, we want a risk analysis task, and subsequent improvement actions, that will not overwhelm the organization. We want to approach risk mitigation with a scalpel, not a shotgun. A careful choice will result in the greatest benefit. Also, remember that we will be approaching social responsibility with a continuous improvement philosophy.

This risk assessment shouldn't be the last risk assessment on this process. It's just one iteration. By choosing the value function(s), we're choosing what we'll work on *first*.

4.5.2 TASK

There are two primary factors to consider in order to choose which value functions will be carried to the FMEA and RA. First, go back to the narrative. Compare the narrative and the value/function diagram. Are there easy-to-identify value functions that provide specific benefits to analyze. For example, if the narrative speaks to a need to reduce hazardous waste, perhaps the value function of "creation of financial reports" might not be the first choice for analysis. Make sure that there is a fit between the value function chosen and the goals identified in the narrative.

Second, use the "voting" visual cue from the value/function diagram. If the team could easily brainstorm many failure modes for a certain value function, that value function might be a fruitful choice for analysis. In order of priorities, making some improvement on a very risky process step may have more impact than making a lot of improvements on a less risky process step.

In summary, choose only the number of value functions upon which the team feels confident in being able to take improvement actions. This will depend on the resources and time available in the organization. Then, use the narrative and the function failure mode "stack up" to determine the value functions with the most impact toward the goals identified. The risk of choosing too many value functions is that the team may get bogged down by an overwhelming amount of work. The risk of choosing too few value functions is missed performance improvement opportunity.

4.5.3 CAUTIONS

The following list describes some cautions for this fourth step of the SRFMEA process, value function selection. Attending to these issues will ensure the team has just the right amount of work ahead.

- Don't forget to refer back to the narrative
- Make sure stakeholders and process experts are involved in the function failure mode brainstorming
- Don't choose more than your resources can handle
- Avoid scope creep by using the process boundaries identified in your SIPOS

4.6 STEP 5: THE FAILURE MODE EFFECTS AND ANALYSIS

4.6.1 REASON

Now we're into the meat of the SRFMEA. The primary purpose of the FMEA is to identify the potential failures and their effects for the function(s) selected in Step 4. If you are analyzing several functions, then you will have several, separate

FMEAs. The FMEA is a prioritization matrix documented as a spreadsheet; an example is shown in Figure 4.6. Use the spreadsheet software most familiar to your organization. For complex functions these can become very large, cumbersome spreadsheets. In large organizations with many cross-functional processes there may be a desire to link FMEAs across functions. These complexities may substantiate the need to purchase software dedicated to FMEA. Such software products exist; seek them out if you need them. Ultimately the team's time needs to be spent on the content of the FMEA, not the management of the spreadsheet.

We are suggesting that your team always consider all seven categories of the ISO 26000 guideline on social responsibility. Even for functions that appear not to touch one of the categories, think hard. There may be extenuating effects. Now is a good time to utilize the pessimists on your team. The generation of potential failure modes is a brainstorming exercise. All reasonable, possible failure modes, for all seven core subjects, will be added to the FMEA. Sometimes it is difficult for the optimists in the room to recognize *potential* failure modes, if that failure has never happened before. People who are naturally risk averse are a great asset now.

One of the difficulties of completing an FMEA is keeping the failure modes and the effects separated through the process. In our experience, this is best accomplished by first brainstorming all possible failure modes, before documenting any of the effects. This is tough to do; we are naturally drawn to claim what bad outcomes could happen—that's an *effect*. What we really want first is to understand how the function can go askew—even with the best of intentions how can things go wrong (the failure mode). Only after we have identified all of these failure modes do we then think of what the impact to society will be (the effect). After all failure modes and effects are listed, then circle back to the modes. Are there additional causes of function failure to create the same effects? This will be an iterative process: mode-effect-mode-effect-mode-effect.

The team may choose to work category by category. For example, in Figure 4.6, we've shown an FMEA for the function of the provision of office equipment. The team may choose to first tackle all modes and effects for human rights, then all modes and effects for environment, for example, continuing until all core subjects are covered. With this approach the team would first brainstorm all human rights failures that could be caused by the provisioning of office equipment. One potential failure is that we purchase our equipment from vendors who do not share our desire, based on the narrative, for consideration of human rights.

We need to recognize here that the definition of human rights in England may be very different than the definition of human rights in Borneo. The operational definition of *human rights* comes from the narrative. This should reflect the organization's desire for continuous improvement in its performance toward human rights. The organization in Borneo may have a very different starting place than the organization in England. That's OK; not everyone has the same standard at the same time. This can get very complicated with global organizations. But we all need to be on a journey of improvement.

If human rights is not a focal point, again from the narrative, the team may choose to do no FMEA analysis on this category and skip right to the next category, environment. Can you see how the narrative is so important? The whole process of the SRFMEA is a

Step 5: Failure mode and effects analysis

Function	Machinery provided for office use: copiers, fax, computers, cell phones, PDAs, desk phones		
Core Subject	**Social Responsibility Failure Modes and Causes**	**Failure Effects**	**Severity Class**
Organizational governance	No known impact		
Human rights	Machinery purchased from supplier nation with human rights issues	Unknown complicity with human rights violations through purchasing practices	8
Labor practices	Inequality between employees in type of device provided	Race, gender, age, etc., discriminatory practice perpetuated	3
Environment	Device fails to function due to defects	Scrapped device in waste stream	9
	Device is not energy efficient	Excess energy used	5
	Device becomes technologically obsolete	Scrapped device in waste stream	9
	Device generates waste during regular operation	Excess paper in waste stream	5
		Excess paper manufactured	6
		Excess paper transported	6
Fair operating practices	No known impact		
Consumer issues	Unreliable equipment causes service disruption	Safety-critical services are disrupted	6
Community involvement and development	No known impact		

FIGURE 4.6   The FMEA.

Failure Mode Effects Severity Table

Category	Question	1	2	3	4	5	6	7	8	9
Fair operating practices	Will this failure have an effect on the industry at large?	No industry impact	Random industry impact	Low known industry impact	Low unknown industry impact	Moderate known industry impact	Moderate unknown industry impact	High known industry impact	High unknown industry impact	Severe impact on industry
Labor practices	Will this failure negatively impact fair labor?	No impact to fair labor	Random fair labor issues	Low known fair labor issues	Low unknown fair labor issues	Moderate known fair labor issues	Moderate unknown fair labor issues	High known fair labor issues	High unknown fair labor issues	Severe impact on fair labor
Human rights	Will this failure cause humans freedom?	No impact to human rights	Random human rights issues	Low known human rights issues	Low unknown human rights issues	Moderate known human rights issues	Moderate unknown human rights issues	High known human rights issues	High unknown human rights issues	Severe impact on human rights
Environment	Will this failure have an impact on the natural environment?	No environment impact	Random environment impact	Low known environment impact	Low unknown environment impact	Moderate known environment impact	Moderate unknown environment impact	High known environment impact	High unknown environment impact	Severe impact on environment
Organizational governance	Will this failure affect governance (reporting) structure?	No impact to governance	Random governance issues	Low known governance issues	Low unknown governance issues	Moderate known governance issues	Moderate unknown governance issues	High known governance issues	High unknown governance issues	Severe impact on governance
Consumer issues	Will this failure cause safety, health, or damage to the consumer?	No impact to the consumer	Random impact to the consumer	Low known consumer impact	Low unknown consumer impact	Moderate known consumer impact	Moderate unknown consumer impact	High known consumer impact	High unknown consumer impact	Severe impact on consumer
Community involvement and development	Will this failure have an impact on society at large?	No societal impact	Random societal impact	Low known social impact	Low unknown social impact	Moderate known social impact	Moderate unknown social impact	High known social impact	High unknown social impact	Severe impact on society

FIGURE 4.6 (Continued).

refinement of focus based on priority. The FMEA will focus our attention by prioritizing the effects of our function's failures in the social responsibility categories of interest as guided by the narrative. This prevents the organization from erring either on the "light" side, by only completing ineffective social responsibility improvements that may receive a lot of publicity but not really make any progress, or on the opposite, "heavy" side and getting bogged down in analysis paralysis, unable to make any improvement because the task is too overwhelming. Using the FMEA as a prioritization matrix highlights the few critical effects that need immediate attention. This is the reason for the FMEA.

4.6.2 TASK

The task of completing the FMEA starts with assembling the team and assembling the spreadsheet. Again, the team should be comprised of process experts, stakeholders, "fresh eyes," and pessimists. This is a team brainstorming activity. We suggest projecting the electronic spreadsheet during the brainstorming activity to avoid the duplication of effort of someone having to retype the results. Take a look at the FMEAs of other similar functions. Sharing of these electronic spreadsheets across the organization may improve the quality of subsequent FMEAs. The team should also be very familiar with the SIPOS and the narrative, before the brainstorming begins.

Choose which of the seven categories, or core subjects, will be excluded from analysis, if any. Then start the brainstorming of the failure modes. A failure mode is an error, glitch, fault, or breakdown in the function. Effects are the outcome of the failure—what society experiences because of the failure. At this time it will be quite natural for the team to gravitate toward identifying effects, not failure mode. Each time an effect is shouted out, direct the team members to identify the failure mode that could have caused that effect. At first, focus on documenting failure modes only. Try to get as many reasonable, possible failure modes for each category. You may go category by category, or mix them all together.

Then begin the identification of effects for each failure mode identified. Most of the time there will be more than one effect from a single failure mode, but sometimes there is only one effect for a failure mode. Now is the time to think about the whole impact to society of your failure mode. This may cross the seven categories. Think long term, think holistically, think about all stakeholder groups. Some of the effects may be minute, some may be horrendous. List them all.

Now start the iteration. Did any of your effects generate new ideas for failure modes? Maybe in a different category? Go back and add the new failure modes. Then complete the effects for those additions. Now go back and forth and back and forth. We want to be exhaustive, within reason. Don't get stuck in the "well if a butterfly flaps its wings ..." scenarios. But don't be too superficial either. Make sure that you've been very thorough for those elements noted in the narrative.

Next is the prioritization. There are two different ways to accomplish this task. This first is to rate effects, and you're always rating effects, not failure modes, across the organization. We've offered an example of an organization-wide matrix of rating for each of the seven categories in Figure 4.6. In this situation, the organization will be concerned with the comparison of different FMEAs. This may be important if the organization expects to have limited resources for the improvement action and wants to then prioritize effects across functions in order to ration the improvement resources. The second method is to rank the effects within a single FMEA. The team gives the most severe effect the rating of 9, and the least effect the rating of 1, and then goes through the process of intra-FMEA comparison of all effects, rating each 1 through 9. With this method, there should never be comparisons of one FMEA to another; the functions themselves may have very different impacts on society. Choose the method most appropriate for your organization. The benefit of the former is an organization-wide holistic approach, but it will be more difficult to rate the effects, and stringent operational definitions for ratings must be documented. The benefit of the latter is a very easy process of rating, but a stand-alone prioritization when completed.

That's it. Now the FMEA is done. Our suggestion is let it sit for a week or so. Give the team some time to process the outcome. Meet again after that week, and make sure the team is still satisfied with it. Add more information if needed. Then put it back to bed.

4.6.3   CAUTION

The following list describes some cautions for this fifth step of the SRFMEA process, the FMEA. Avoiding these issues will ensure the team has an exhaustive list of possible failure modes and has carefully considered the effects of these failures.

- Having only optimists on the team
- Thinking only near term for effects
- Getting off track—stay focused by using the narrative
- Mixing failure modes and effects
- Not conscientiously planning the appropriate rating technique
- Not being exhaustive and iterative with brainstorming

4.7   STEP 6: RESPONSIBILITY ANALYSIS

4.7.1   REASON

In the AIAG method of FMEA, our Steps 5 and 6 are completed at the same time. We advocate separating the prioritization of the effects and the analysis of the frequency of occurrence of the effects into two separate steps, first the FMEA and then a separate RA, like the original MIL-STD-1629 directs (we are replacing the military standard nomenclature of *criticality analysis* with *responsibility analysis*). Our experience has been that combining these steps leads to confusion. The FMEA becomes an overly onerous chore, thus leading to FMEA as paper on a shelf, not a living document. We've shown this separated step in Figure 4.7.

In the AIAG method for completing FMEA, the risk priority number (RPN), which creates the prioritization for action, is calculated through a mathematical manipulation of severity class, occurrence class, and detection class. In MIL-STD-1629, in

Step 6: Responsibility analysis

Function	Machinery provided for office use: copiers, fax, computers, cell phones, PDAs, desk phones					
Core Subject	Social Responsibility Failure Modes and Causes	Failure Effects	Severity Class	Occurrence Probability	Responsibility Priority Number	Remarks and Justifications
Fair operating practices	No known impact					
Labor practices	Inequality between employees in type of device provided	Race, gender, age, etc., discriminatory practice perpetuated	3	1	31	Current job analysis determines personal device needs for all employees
Human rights	Machinery purchased from supplier nation with human rights issues	Unknown complicity with human rights violations through purchasing practices	8	9	89	No current practice to screen suppliers for human rights issues
Environment	Device fails to function due to defects	Scrapped device in waste stream	9	5	95	All devices currently sent to a recycler; however, recycler's practices have not been audited
	Device is not energy efficient	Excess energy used	5	2	52	Government-certified high-energy products are currently required by purchasing procedures

FIGURE 4.7 The responsibility analysis.

Device becomes technologically obsolete	Scrapped device in waste stream	9	5	95	All devices currently sent to a recycler; however, recycler's practices have not been audited
Device generates waste during regular operation	Excess paper in waste stream	5	9	5	No policy communicated or monitored for minimizing printing
	Excess paper manufactured	6	9	69	
	Excess paper transported	6	9	69	
Organizational governance	No known impact				
Consumer issues	Unreliable equipment causes service disruption	Safety-critical services are disrupted	6	1	61	Triple redundant call center systems are currently in place
Community involvement and development	No known impact				

FIGURE 4.7 (Continued)

the criticality analysis, there is no mathematical manipulation to create an RPN, but rather a graphical, qualitative comparison of occurrence and severity is used. We're suggesting a morphing of these two methods. Our responsibility analysis will generate an RPN, as a compound number, based upon severity and occurrence. This will identify the priority for action.

We eliminate the AIAG FMEA concept of detection class. With respect to social responsibility, detection amounts to awareness. With social responsibility, awareness is always shifting. In the 1920s all of society was unaware of the negative impact of lead paint. In the 2000s most of society was unaware of the ultimate, global social impact of subprime mortgages originating in the United States. Because awareness is constantly shifting, detection cannot be a prioritization element. However, increasing awareness may be one action, hopefully not the only action, in the mitigation of the risk of the failure mode.

We suggest a deeper analysis than the graphical categorization of severity and occurrence than MIL-STD-1629. We are suggesting that the severity class rating from the FMEA and the occurrence class from the RA be combined to form a compound number. For example, if a single effect has a severity of 9 and an occurrence of 6, the RPN becomes 96. If a single effect has a severity of 6 and an occurrence of 9, the RPN becomes 69. As can be seen from these two examples, this type of RPN creation has a very different outcome than creating the RPN through the product of the severity class and occurrence class. If the product were used, these two examples would generate the same RPN, 54. It is our contention that our first priority should be to eliminate effects, rather than the rate of occurrence of the effect. The compound number method will allow for the highlighting of higher effects in the responsibility analysis prioritization.

Also, like MIL-STD-1629, we have chosen to emphasize the ability to note probability of occurrence as either a qualitative or quantitative rating. We have shown both sample ratings in Figure 4.8. This demonstrates a very simple, overall, expert-opinion-driven probability of occurrence. If the probability of occurrence is known, then use the quantitative rating. These two methods could even be combined in the same RA. For example, if the probability of human rights violations is not known as an actuarial statistic, then a qualitative probability can be assigned by the team. If, in the same RA, the quantitative data for probability of human trafficking are known, then the quantitative table can be used to assign the occurrence class. The point is prioritization. Which are the most probable occurrences? Which are the least probable occurrences? That's all we need to know.

4.7.2 TASK

In order to complete the responsibility analysis the team should be familiar with the SIPOS, the narrative, and the FMEA. It is not necessary that the exact same team be assembled for the responsibility analysis. The pessimists may leave the room; however, the statisticians and the actuaries may now be invited. Stakeholder representation and process experts are still needed. Start a new spreadsheet, for this additional, and final, prioritization. The first four columns of the responsibility analysis are carryovers from the FMEA (Figure 4.6).

Failure Mode Effects Occurrence Table

A. Quantitative Occurrence Table

Occurrence	Definition	Occurrence Probability Level
Frequent	Greater than 20% probability of failure	9
Reasonably probable	Greater than 10% probability of failure	7
Occasional	Greater than 1% probability of failure	5
Remote	Greater than .1% probability of failure	3
Extremely unlikely	Less than .1% probability of failure	1

B. Qualitative Occurrence Table

Occurrence	Definition	Occurrence Probability Level
Frequent	A high probability of occurrence; probable failure during most process cycles	9
Reasonably probable	A moderate probability of occurrence; high likelihood of a single cycle failure	7
Occasional	An occasional probability of occurrence; some likelihood of a failure during at lease one process cycle	5
Remote	An unlikely probability of occurrence; remote probability of failure during a single process cycle	3
Extremely unlikely	The probability of failure is almost zero	1

FIGURE 4.8 Quantitative and qualitative occurrence ratings.

Using either the qualitative, quantitative, or combination methods, rate each effect for its probability of occurrence. Again, the ratings are from 1 to 9. Don't waste time arguing a rating of 5 versus 7. Borrow data from other RAs if they are available. If SRFMEAs will be compared across the organization, make sure qualitative ratings are consistent across the organization for the same effects. If the RA will stand alone, and never be compared, simply rate the most probable to the least probable for all effects. In either case, the organization must define the frequency, either qualitative or quantitative, for each occurrence rating.

Then create the RPN by producing a compound two-digit number from the severity class (digit in the tens place) and the occurrence class (digit in the ones place). This yields the lowest possible RPN of 11, and the highest possible RPN of 99. Take a look at the RPNs. Do they make sense? Sometimes, if the ratings took place in exhaustive hours-long sessions, there can be rating creep in either too lenient or too stringent directions. After all RPNs are generated, do a sanity check to ensure that they make sense. Making small tweaks is OK at this time. Just make sure organization-wide changes to severity or occurrence tables reflect the knowledge generated from the sanity check.

The last column of the RA is just as critical as the RPN. Don't shortchange it. These are the remarks and justifications. This is the description of "How did we get here? How did we get into this mess?" as reflected in the current state that yielded the RPN. There are resource, knowledge, time, regulatory, and other constraints that caused the current state. These are known by the process experts; get them documented on the RA. You will have to address these same constraints when you start your improvement actions. These aren't excuses for not taking action; they are just obstacles that will need to be challenged when improving.

This completes the responsibility analysis. Leave the document alone and let it sit for a week or so. Reconvene the team to review. Make any adjustments and add any remarks. Let it sit again. Allow some time for the team members to mentally process the RA results before moving on to the improvement actions.

4.7.3 Cautions

The following list describes some cautions for this sixth step of the SRFMEA process, the RA. Avoiding these issues will ensure the team has identified the most important effects upon which to take action.

- Getting stuck arguing over insignificant differences in occurrence ratings
- Skipping the remarks and justifications
- Failing to prioritize—everything is a 5
- Not doing a sanity check for scope creep
- Spending too much time or too many resources on seeking exact quantitative occurrence data

4.8   STEP 7: ACTION PLANNING AND TRACKING

4.8.1 Reason

Now the goal is in sight. We're ready to identify actions to mitigate risks to social responsibility. Can an organization skip straight to improvement actions without the analytic steps of the FMEA and the RA? Sure. But how will you know if your effort has the right impact? Or any impact at all? Approaching improvement actions after careful analysis of potential risks, severity, and occurrence helps the organization focus minimal resources on improvements for maximum effect. Without this analysis your organization could be applying the maximum resources for a minimal effect.

Your RPN tells you which effects to act on. Some organizations choose a cutoff RPN to mandate improvement. For example, all effects with an RPN over 84 require improvement. We strongly encourage you to *not* fall into this trap. For one, this RPN may be an arbitrary number applicable only to this function. An 84 on one SRFMEA may be just as important as a 68 on another SRFMEA. Our experience with organizations requiring action at an arbitrary cutoff creates a "gaming" of the severity class and occurrence class. Teams may begin to underestimate the rating in order to keep the RPN below the action-required threshold.

A better approach is to reflect on the resources available for corrective action. Pick the top 3 or 5 or 10 RPNs for action based upon the resources available. For some small organizations this may only be one action at a time. That's OK. Remember, we're after small, steady improvement, continually, over time. On the other hand, the team may pick the top three RPNs and find that some sort of process change is already planned for those effects. The team may then decide to pick up the fourth and fifth priorities, but skip the first and third. Use a combination of the team's best judgment, the narrative, organizational resource constraints, and the RPN to select the total number of effects upon which to focus improvement. In the example in Figure 4.9, the team has chosen the top RPN in each of two different categories: human rights and the environment. Our advice is to pick just a few and do them well.

When generating ideas for the improvement actions, the best solutions address the failure mode. The best improvement eliminates the possibility of the failure mode. Second best is the reduction of the severity of the failure. Third best is to reduce the probability of occurrence of the effect. And lastly is simply to raise awareness on the effect. Perhaps your actions will be a combination of these types of solutions. A check on the effectiveness of your improvement action is its ability to significantly reduce the RPN of the effect.

Actions are only effective when they're completed. Effective action plans identify the action, the person responsible for the action, and the timing expected for the completion of the action. There should be accountability and visibility of organizational leadership interest in the improvement actions. The actions from many SRFMEA efforts may be combined into the organization's improvement initiatives. Whether your organization uses root cause problem solving, plan-do-check-act, Six Sigma, Lean Production, kaizen, or other continuous improvement initiatives, make sure that your social responsibility improvements are an integral element to the same programs, not an add-on. For example, high RPN social responsibility risks may become project ideas for your Six Sigma program.

In summary, the reason for action planning and tracking is to ensure that mitigation of the risks identified through all of the SRFMEA work is corrected. This is the crux of the whole effort. Our goal is not only to identify process weaknesses that may lead to irresponsible behaviors, but also to build robust systems that ensure responsible behavior. Effective project management of the corrective actions is just as important, if not more important, than any of the previous steps.

4.8.2 TASK

Begin the third spreadsheet of the SRFMEA process, as shown in Figure 4.9. Transfer the first four columns from the RA to this new action planning spreadsheet, for only those effects chosen for improvement. Reducing the RPN of a few effects, and doing it well, is much more effective in the long run than trying to tackle too many actions at once and losing focus with your resources.

Use creativity and innovation techniques to generate improvement ideas for each of the effects chosen. Generate many possible solutions to improving the effect,

Step 7: Improvement actions and tracking

| Function | Machinery provided for office use: copiers, fax, computers, cell phones, PDAs, desk phones | | | | | |
Core Subject	Social Responsibility Failure Modes and Causes	Failure Effects	Remarks and Justifications	Improvement Plans	Assigned To	Due Date
Human rights	Machinery purchased from supplier nation with human rights issues	Unknown complicity with human rights violations through purchasing practices	No current practice to screen suppliers for human rights issues	1. Train all purchasing employees on human rights violations and how to detect 2. Require on-site supplier audits for the detection of incidents of discrimination, forced or compulsory labor, child labor, freedom for collective bargaining, and known previous human rights violations	1. J. Morganstern 2. J. Brink	1. July 15 2. August 30
Environment	Device generates waste during regular operation	Excess paper in waste stream Excess paper manufactured Excess paper transported	No policy communicated or monitored for minimizing printing	1. Provide training in electronic reading and note taking 2. Develop awareness campaign posters to be posted at all printers 3. Audit all printers for parameter settings to eliminate unneeded banners, etc. 4. Review all management reporting requirements to eliminate any unnecessary printing requirements 5. Outfit all conference rooms with electronic media access and projection	1. M. Madsen 2. K. Smith 3. J. Beeman 4. F. Buelman 5. T. Jones	1. August 30 2. September 7 3. September 7 4. December 31 5. December 31

FIGURE 4.9   Example: Improvement actions and tracking.

and then use analytic, experimental, cost-benefit, and decision-making methods to choose the best solution from among many possible solutions.

Once solutions have been chosen, identify the tasks needed to implement the solutions. Identify the single person responsible for ensuring that each work task is completed, when the work task will begin and end, and how that work task will be validated as complete. Make sure there is a system of management oversight and reporting.

If the organization has completed the top few selected items, and has resources to apply to the next round of RPNs, start the implementation planning process again. Once the process of SRFMEA has been completed, this analysis could drive continuous improvement priorities for months to come. Always go back to the FMEA and RA and revise the RPNs as solutions are implemented. Your new prioritization for the next actions is already ready for you.

4.8.3 CAUTIONS

The following list describes some cautions for this seventh step of the SRFMEA process, the action planning and tracking. Avoiding these issues will ensure the team accomplishes performance improvement. Effective action planning and tracking will ensure that the most improvement happens in the least amount of time.

- Failing to identify the single person responsible for each implementation task
- Identifying too many improvement actions
- Not revising the RPNs and failing to continue to move forward after the completion of the first actions
- Failing to involve the stakeholders in the improvement actions
- Not utilizing the SRFMEA as a living document

5 Using SRFMEA to Improve Social Responsibility Practices and Processes

What to expect:

- SRFMEA for social responsibility improvements
- SRFMEA stakeholder importance
- SRFMEA mission
- SRFMEA goals
- Actions after the SRFMEA
- Prioritization of SRFMEA actions

5.1 IMPROVEMENTS BEYOND THE SRFMEA

The ultimate goal is not a library full of SRFMEA risk assessments. The ultimate goal is a socially responsible organization, an organization that is continually striving for improved performance. Even more important than recognizing the opportunity to improve social responsibility performance is achieving it. This chapter discusses what to do to improve performance after the risk assessment is complete.

The goal of the SRFMEA is not the failure mode analysis itself. The goal is the action that comes from the completion of the process, and the culture that results from improved awareness. The analysis of suppliers and customers, and inputs and outputs, improves the organization's understanding of its role in the larger community. The development of the narrative tells the story of the vision toward which members of the organization strive. Detailed process flow analysis identifies easy opportunities to eliminate non-value-added tasks and helps to focus the improvement effort. Choosing a function upon which to focus rallies organizational resources around important causes. All of these critical steps in the FMEA processes only lead to the risk analysis. And completing the risk analysis does not necessarily result in process improvements.

The FMEA itself initiates the evaluation of each value/failure diagram against the different core subjects of social responsibility. Awareness of risk is quantified through the assignment of severity class. Then the accompanying responsibility analysis polishes the prioritization of risk by applying the occurrence class. A combined high severity and high occurrence brings focus to those urgent and important needs. The identified actions, projects, or programs that offer the avenue for improving social responsibility, the whole organization, process upon process, function upon function, by completing just a few remedial actions for each SRFMEA, moves forward, kaizen style, to achieve continual improvement of social responsibility.

This continuous improvement is the goal of the SRFMEA. Recognizing that sustained committed effort in every process, all of the time, is the way to achieve social responsibility. The SRFMEA is only the road map; the continual improvement of the process is the journey. This journey requires listening to stakeholders, being aware of changing knowledge and paradigms of what is responsible behavior, mindfulness in planned actions, and a striving to analyze risk and take corrective action. The SRFMEA process is the path by which the journey is made. While we believe that completing the SRFMEA identifies and prioritizes actions for socially responsible behavior, there are many other quality tools that can be utilized in concert with the SRFMEA. We will discuss a few of these other tools, and their potential benefits, in a social responsibility improvement program.

5.1.1 SOCIAL RESPONSIBILITY CONTINUOUS IMPROVEMENT GOALS IN SUMMARY

The ultimate goal of the SRFMEA is a tool, not only for risk mitigation, stakeholder satisfaction, or market advantage, but also as a mechanism for the instillation of a continuous improvement culture toward social responsibility. Social responsibility is an ideal, ever changing, ever pressured by shifting paradigms. Immediately after the SRFMEA is complete and risk priority numbers (RPNs) reduced, it's time to go again. Individual involvement throughout the organization, to seek this ideal through continuous improvement, should be the result desired. At each step of the SRFMEA process, SIPOS, narrative, value function analysis, failure mode analysis, responsibility analysis, and continuous improvement action help to build this culture. Awareness of risk is increased, connection to stakeholder is solidified, and organizational culture is made. Social responsibility continuous improvement is a journey, not a state; it is a way to think, not a thing to do; it is an ideal, not a product or service feature.

5.2   AFTER THE SRFMEA

While the process proposed in this text results in an SRFMEA risk priority number (RPN), the resulting RPN might not be the optimum stopping point. Building a culture of continuous improvement is a difficult balancing act. We want to deploy tools, like the SRFMEA, which spark the potential for myriad improvements. But at the same time, it is easy to overwhelm the organization with too much activity; resources become inundated and ineffective, too many processes are modified concurrently, and we lose control. Therefore, additional tools might be needed in order to rank the projects resulting from the SRFMEA. This section will review some of these additional options.

5.2.1   IMPROVEMENT PRIORITIZATION

After completing the SRFMEA, the resulting RPN provides a ranking of projects to be initiated. We could look at any completed responsibility analysis (RA) and immediately be able to identify the failure risk with the largest magnitude of potential loss and the most probable to occur. Then improvement projects are launched based on the severity class and occurrence class of a particular social responsibility failure effect. If the SRFMEA RPN is utilized, the project ranking and implementation will be based on the highest RPN. If there are other organizational concerns that need to be addressed beyond this categorization, such as cost, then additional ranking methodologies can be utilized. If using RPN to prioritize actions, this is an ideal stopping point for addressing social responsibility risks; however, we want to explain a few more methods of prioritization. In the following sections we will cover some tools that may help your organization create the balance of attacking the most important opportunities without overwhelming resources. These tools include plan-do-check-act, the prioritization matrix, risk-cost-benefit analysis, multicriteria decision making, the decision tree, and the network linkage model. Each of these methods is being presented to help with the prioritization of projects after the SRFMEA has identified the most severe and probable risk to socially irresponsible behavior, found through the RPN.

5.2.2   PLAN-DO-CHECK-ACT

Plan-do-check-act (PDCA) is a well-known and commonly used approach to continuous improvement. PDCA was developed by Shewhart in the 1930s, and later made more popular by Deming. Deming, in his interpretation, changed *check* to *study*, or PDSA. The steps in the PDCA process involve planning for a change (*plan*), implementation of the change (*do*), reviewing the results and lessons learned (*check/study*) of the change, and taking action to adopt the change (*act*) or repeat the cycle. This continuous repetition of identifying improvement opportunities, evaluation, and implementation is a process that can be utilized in any scenario.

In the context of the SRFMEA, PDCA can be utilized once the analysis is complete. For example, a company may identify that it wants to improve on its level of air pollution emissions. The PDCA process can be interjected to identify a change for reducing emissions (*plan*). The implementation of a new exhaust system to better filter discharges from the process (*do*) is undertaken. The emission levels before the change and after the change are monitored to see if the new filter reduces emissions (*check/study*). Based on the successful results of the implementation, the company decides to implement similar filters at its other plants with similar processes (*act*). This process can be repeated over and over, in the essence of a true continuous improvement process.

5.2.3   PRIORITIZATION MATRIX

The prioritization matrix is a very simple tool often used in a team setting, with everyone's opinions contributing to the completion of the prioritization. Figure 5.1 shows a completed example. For many of our identified improvements there are competing factors of consideration when deciding on the improvement projects to begin. In this example, the cost of implementation of an improvement is compared to the alignment with corporate strategy. The money we have to invest in solutions is limited. And some ideas contribute better to corporate strategy than others. In this simple example, we place rank each solution idea in blocks as either high, medium, or low cost of implementation. And then each idea is ranked toward a second factor, the linkage to corporate strategy. This is a simple two-factor prioritization matrix. The cost and linkage to strategy are considered evenly weighted toward a decision to move ahead with the project to implement the solution. Opportunities that are low cost and high linkage to corporate strategy should be implemented first. When faced with multiple opportunities, think about the factors that are most important to the decision-making process. Complete a simple prioritization matrix to identify the most lucrative opportunities. If, in addition to the ranking by RPN, the company is concerned with the linkage to corporate strategy and the cost of implementation, the resulting projects from the SRFMEA can be placed in a prioritization matrix based on these two criteria.

5.2.4   RISK-COST-BENEFIT ANALYSIS

A cost-benefit analysis can be conducted as well using the SRFMEA RPN ranking. A failure mode, identified with the highest RPN, it can be analyzed based on intervention

		Link to Corporate Strategy (be a more socially responsible citizen)			Action Items
		High	Medium	Low	
Cost of Implementation	Low	1. Add new filters to exhausts to reduce harmful emissions			1. Add new filters to exhausts to reduce harmful emissions
	Medium				2. Tear down and rebuild portion of the warehouse to better manage cardboard waste
	High	2. Tear down and rebuild portion of the warehouse to better manage cardboard waste		3. Implement more cost-efficient equipment throughout the plant and replicate in the other 10 plants	3. Implement more cost-efficient equipment throughout the plant and replicate in the other 10 plants

FIGURE 5.1   The prioritization matrix.

and cost-effectiveness data. For example, Wilson and Crouch (2001) provide a listing of 500 lifesaving interventions and their cost-effectiveness based on 1993 U.S. dollars of the cost/life-year of each intervention. Current values for the interventions associated with identified risks can be calculated in current year costs. If, for example, one of your highest RPN values is related to asbestos and the intervention selected is to ban asbestos from sealant tape used in your process, the estimated cost/life-year (in 1993 costs) is $49,000,000. These values from tables can be used to supplement the SRFMEA RPN and determine a suitable ranking for project selection and implementation.

5.2.5 Multicriteria Decision Making (MCDM)

MCDM is a decision-making technique that involves identifying all of the relevant criteria important to the decision process. It's a lot more complex than the simple prioritization matrix. If there are multiple criteria (three or more), then this is an appropriate supplemental technique for choosing the best opportunity for improvement. The multiple-criteria scoring model involves listing all of the criteria that you want to evaluate for each project and then weighting each of the criteria. The weighting and ranking of the criteria result in a weighted score that is used to select the project order for implementation. In the example in Figure 5.2, the company wants to select its social responsibility projects based on SRFMEA results and organizational elements such as cost, link to strategy, and global impact. The three different SRFMEA results, represented by RPN 1, RPN 2, and RPN 3, are analyzed against the three different factors of importance, cost, link to strategy, and global impact. With the MCDM technique, many factors of importance can be simultaneously rated. Each of the three organizational factors is weighted based on importance to the firm, and then scored a 1, 3, or 5. In the example shown, we see that the solutions presented for the failure mode represented by RPN 3 is the best for addressing the three factors of concern. With limited resources to devote to improvement projects, this organization should choose to focus on improving this single failure mode, first, even though the RPN may not be the highest of the three.

Criteria	Score For			Criteria Weight
	RPN 1	RPN 2	RPN 3	
Cost	3	1	5	0.4
Link to Strategy	1	5	3	0.3
Global Impact	1	3	5	0.3
Weighted average	1.8	2.8	4.4	1

FIGURE 5.2   The multicriteria decision-making tool.

5.2.6 DECISION TREE

Yet another tool to consider when trying to juggle many competing considerations is the decision tree. The decision tree is utilized when there are multiple scenarios that one might want to consider in the decision-making process. This is a more linear approach than the MCDM. Rather than considering simultaneously competing factors in the decision, the decision tree analyzes the potential sequential factors of importance. The tree is a representation of the problem with nodes and branches, with each branch considering an "if" this decision, "then" this outcome arises. The nodes reflect a decision point and are represented by squares and circles. The branches represent the alternatives that one has at each decision point. If you wanted to make a decision on which improvement project to implement first, a decision tree could be utilized, as shown in Figure 5.3.

5.2.7 NETWORK LINKAGE MODEL

Many of the improvement projects focused on changing socially responsible behavior will need to change the behaviors of many different stakeholders in different ways. And a lot of this behavior is linked in a web on interconnected action and reaction. A network linkage model involves evaluating all of the important risks associated with the biggest opportunities. An example of this is shown in Figure 5.4. It is a graphical technique that displays the interconnectedness between stakeholders and their actions. Opportunities that create improvements for many stakeholders, in multiple ways, may be prioritized over those with limited reach or benefit. The network linkage model can also help the improvement project team recognize the complexity needed, toward stakeholder engagement, in the solutions to be implemented.

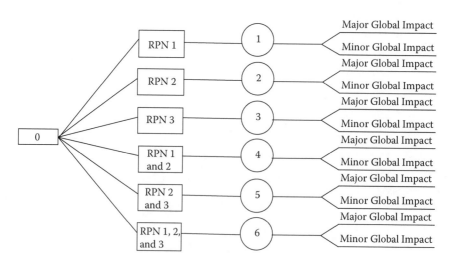

FIGURE 5.3 The decision tree.

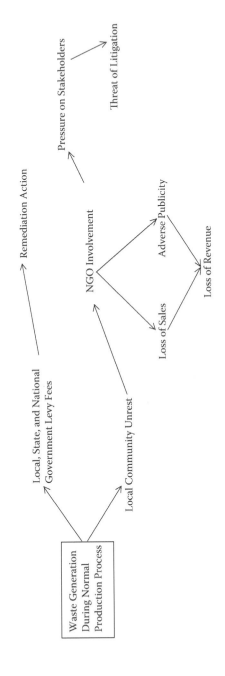

FIGURE 5.4   The network linkage model.

5.2.8   RISK PROFILE MAP

The final technique that we would like to offer as a supplement to ranking RPNs when choosing the focus of improvement is a risk profile map. A risk profile map allows you to take the RPN values and map them in order to get a visual representation of where they might fall relative to important elements within the company. The example shown in Figure 5.5 demonstrates similar information from the prioritization matrix discussed earlier. In this example, the risk profile map shows the relative distance, in the dimensions of cost and linkage to corporate strategy, for four different failure modes labeled by their RPN number. Those opportunities that land in the upper, left quadrant in this example would be preferred. These failure modes, if improved, would have a low cost of improvement, and a high linkage to the corporate strategy.

5.2.9   PRIORITIZATION IN SUMMARY

Rarely does an organization have all the resources to simultaneously attack its highest social responsibility risk mitigation opportunities. We have to pick and choose our battles. Ultimately, we are talking about changing behaviors. Initiatives to target changing behaviors should be carefully deployed. Many simultaneous behavior changes, by different individuals and groups, can quickly become counterproductive to the intended outcomes. Therefore, even after the SRFMEA has done its job of prioritizing risks, we need to be thoughtful and careful about the improvement initiatives we begin. We have demonstrated just a few prioritization methods that could be used for this purpose. Whatever tool your organization chooses to use, be quick about it. Choose the best opportunity for now, get the improvements implemented,

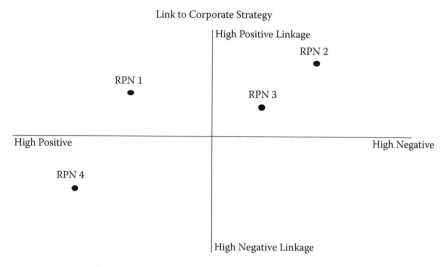

FIGURE 5.5   The risk profile map.

and go right back to the prioritization process to start the next improvement. This is the culture of continuous improvement: careful, controlled change for improvement, completed over and over again, without pause.

5.3   MORE COMPARISON TO THE QUALITY MOVEMENT

We have compared social responsibility to the quality movement. Prior to the quality movement quality was perceived as a state. A product was either of good quality or poor quality. We now recognize that what may be defined as the best quality today may rapidly change to poor quality in the future. We now recognize that quality is not a thing, or a state, or a feature, but an ideal. We strive for better and better quality. What is acceptable today will be surpassed by the competition tomorrow. Nor will we ever actually achieve perfect quality. Social responsibility is also an ideal, not a thing or a state or a feature. It's never actually achieved. Our awareness is always increasing. What is seen as responsible today will be perceived as ignorantly abhorrent at some future time.

Also, like the quality movement, social responsibility can become a market differentiator. Being on the forefront of understanding social responsibility risk allows the organization to be the first to recognize and therefore mitigate the risk. When competing in the marketplace with competitors of equal product function, product availability, or product quality, superior social responsibility has been proven to influence consumer decision making.

As Toyota Motor Corporation has famously shown, a decades-long focus on product quality has not only resulted in market domination, but also created an organizational culture that has been the study of desired replication. Entire consulting markets exist for educating clients in the Toyota Production System for companies seeking marketplace advantages in operations and quality. Quality has become an organizational culture. So to is the goal of the SRFMEA. Involvement in the SRFMEA process by everyone in the organization will lead to an organizational culture where awareness and mindfulness of social responsibility risk are embedded in everything the organization does. One of the goals of the SRFMEA process is this organizational culture shift.

The whole organization's gradual improvement toward the ideal of socially responsible behavior becomes the goal and objective. Using SRFMEA as a tool of continuous improvement propels the organization along the path of progress in this journey. As we speak of the continuous improvement of social responsibility behavior, the metric of improved performance can become tangible through the reduction of RPN. We can "see" the progress toward the ideal.

5.4   VOICE OF THE STAKEHOLDER

One cultural shift achieved from dedication to the SRFMEA process is the connection to stakeholders. Like the quality movement resulting in an organizational culture with attention to robustness in product and process quality, the social responsibility continuous improvement movement will result in an organizational culture in which each employee has vivid awareness of the organization's connection to its stakeholders. Recognition that the organization is an integral part of a community is achieved.

Throughout the SRFMEA process this connection to the community of stakeholders is strengthened. The completion of the SIPOS recognizes who these stakeholders are. The narrative describes the needs of these stakeholders. The FMEA evaluates the potential impact to these stakeholders on every social responsibility aspect. The best continuous improvement actions involve the stakeholders in the solution. Awareness of the connection to the community is enhanced.

An organization that regularly listens to its stakeholders gains a mindful culture. Ultimately, mindfulness yields proactiveness. The continuous improvement of social responsibility takes place earlier and earlier in the business and at lower and lower risk of failure. The organizational culture becomes known for its dedication to social responsibility in the community of stakeholders. At this level the organization is proactively improving social responsibility, not just reactively changing compliance to social responsibility.

5.5   CREATING A MISSION

Eventually the organization needs to get down to the business of setting goals and objectives for its social responsibility continuous improvement program. However, no organization can tackle everything at once. The organization should first start with a vision and then a mission. From the very highest levels of the organization, a vision of commitment to acting responsibly in every aspect of social responsibility must be made. What does the future state look like? Why does the organization want this vision? What is compelling this vision? What are the advantages? Then, also from the highest levels of the organization, the mission should be stated. The mission of the social responsibility continuous improvement program should be created in consideration of the specific conditions of the organization. What will be the focus? Which social responsibility aspects are of particular concern? What market conditions will cause obstacles? The mission statement is when the organization gets personal. How will the social responsibility continuous improvement program be tailored to best benefit this organization?

It is only after the mission and vision of the whole organization are developed that any narrative for any single SRFMEA can be written. Although each SRFMEA will have a unique narrative, each narrative should waterfall from the organization's social responsibility mission and vision. The responsibility analysis and its subsequent action plan then become the tactics of social responsibility; the vision and mission statement are the strategy.

One example of organization-wide goals and objectives may be to reduce organization-wide risk priority numbers, with emphasis on either severity class or occurrence class. For example, an organization may set its next year's goal to be a 20% reduction in company-average RPN. Company-average RPN is created by averaging all line item RPNs of all SRFMEAs. The accompanying objective may be to focus on the reduction in severity risks in the painting operations, and the reduction in occurrence risk in the purchasing function. Both of the previous objectives having been based upon a review of the existing SRFMEAs. Of course, the addition of new SRFMEAs should not be included in the monitoring of the improvement metric. This might negatively, or positively, alter the RPN average.

One risk of setting goals and objectives on the risk analysis outcomes, or RPN, is the "gaming" of ratings during the SRFMEA process. If the vice president of operations in Brunei is personally measured against an RPN maximum allowed, the SRFMEA process and assignment of risk may be "watered down." One solution to this risk is to use company-wide operational definitions for severity class and occurrence class. In this example, organizational rewards and recognition would be planned if the RPN reduction is achieved.

The creation of a social responsibility vision, mission, goals, objectives, and narratives is a way to involve the whole organization in social responsibility. The ultimate goal is an organizational culture of continuous improvement toward social responsibility. Involvement is key. Effective narratives lead to effective responsibility analyses. Effective narratives help align resources and provide focus to mitigate the most important issues. Company-level vision and mission for social responsibility, and thoughtful goals and objectives are needed to create these effective narratives.

5.6   BUILDING SOCIALLY RESPONSIBLE CORPORATE CULTURES

Some careful attention to a few precautions will yield a continually improving socially responsible corporate culture. The SRFMEA, as with any type of reliability assessment, can be completed too liberally or too conservatively. Striking a balance between the recognition of all risks, without getting stuck in "analysis paralysis," and wanting to change everything all at once is the goal. The following section is presented to provide specific precautions and techniques to find this balance.

5.6.1   THE SRFMEA IS A PROCESS, NOT A SPREADSHEET

Foremost, a focus on the process of completing the SRFMEA, as opposed to just filling in the blanks on the spreadsheet, is a recipe for a successful risk assessment outcome. Teams that "cut to the chase" and ignore the process steps leading up to the FMEA matrix or the responsibility analysis will end up with an inferior analysis. Each of the steps of the SRFMEA process is critical and should be accomplished with thoroughness and attention to detail. For example, ignoring the SIPOS will result in missing the identification of important stakeholders, process inputs and outputs, and supply chain providers. It is easy for inexperienced teams to assume that all the SRFMEA team members know who the stakeholders are, and what the process inputs and outputs are. However, taking this shortcut can lead to entire categories of process outputs to be missed in the subsequent risk assessment.

For many SRFMEA teams, the development of the narrative is a difficult process. Many teams are filled with members who may find a focus on the implementation of tactics more comfortable. Developing strategy and vision may be foreign activities. However, taking a shortcut in the narrative will lead to an SRFMEA without a focus or goal. This will eventually lead to SRFMEA documents that are not living documents; the lack of focus and goal does not contribute to the effectiveness in the continuous improvement of the organization. Attention to the development of the narrative of the risk assessment ultimately ensures that the SRFMEA becomes an active and engaged element of the organization's achievement of its objectives.

During the SRMEA process, there may be pressure to skip the value/failure diagram. Or, there may be pressure to pull a process flow diagram from some other process library or documentation. If the organization uses these detailed process flow diagrams as a frequent, normal part of operations, then they may be great foundations upon which to begin the SRFMEA. However, the SRFMEA team should avoid situations in which the team assumes that the preexisting process flow diagram is correct for the use of social responsibility risk assessment. The team should practice *genchi gembutsu*, or "go see for yourself." Walk the process. Follow the process. Observe all energy entering and exiting the process; observe all material entering and exiting the process; observe all human factors involved in the process. The action of developing a value/function diagram for an SRFMEA is different than process flow diagrams generated for other purposes. Go to the process, and ensure that the process of interest for the SRFMEA is thoroughly understood, in every detail, with respect to the seven elements of social responsibility.

Only after these important SRFMEA process steps are complete should the team engage in the spreadsheet activities of the FMEA and responsibility analysis. These two actions may be the most fun, the most engaging, or the most interesting, but unless the appropriate prework is complete, the risk analysis may be simply erroneous. A thorough, technical risk assessment in consideration of the wrong stakeholders, without a goal, and without a detailed understanding of the value-added functions of the process, is of no use. In fact, it could be detrimental to an organization.

For example, if the SRFMEA team fails to walk the process to validate the value/function diagram, it may miss the fact that spent cell phone batteries are improperly disposed, overly focus on the country of origin for the purchase of the phone, and miss this more critical aspect of the social responsibility improvement target of hazardous waste disposal. An SRFMEA that skips the SIPOS process step, during the design of a new financial reporting computer system, may miss the recognition of managers as important stakeholders. Later, when excessive printout copies are required of the system for the managers' use, a socially responsible optimization may have been missed. Each step in the process of creating an SRFMEA is important. Do not take shortcuts. Recognize that the output of each process step leads to the input of the next, and a high-quality job on one step leads to the ability to deliver a higher-quality product at the next step.

And one last note on the SRFMEA as a process, not a spreadsheet: The team should recognize that this is a closed-loop process. Upon completion of an SRFMEA, actions for continuous improvement are begun. By its very nature, continuous improvement implies change to the process. A change to the process yields changes to the SIPOS, and the improvement yields changes to the narrative. By implementing continuous improvement, a need for reevaluation in the risk assessment is initiated.

An SRFMEA is never done; it is never filed away in a library. The next level of assessment and improvement is ready to start immediately after the first has been accomplished.

5.6.2   WHO IS INVOLVED IS CRITICAL

Throughout this book we have referred to those people completing the SRFMA as "the team." We mean this in the truest sense—not a group, not an individual, but a team. For us, a team differs from a group in that the success of the team is contingent upon every member of the team being successful. There is an interdependence among all members of a team. During the SRFMEA process there may be individuals on the team who have a better grasp of the stakeholder community, other members who have a mastery knowledge of the product, and still other members who are creative problem solvers. Each of these members has his or her critical role in the ultimate success of the team. A well-done SRFMEA completed with a team that is ignorant of stakeholders, products, or unable to generate solutions will not achieve successful risk mitigation. The overall team success relies on each member of the team fulfilling a critical role, and being successful at that role.

As with any team, dynamics can be highly engaging or problematic. Decision-making processes can often become sources of conflict on teams. It is recommended that an organization begin its team-based SRFMEA process by becoming trained in effective decision-making and conflict resolution methods. We have found that teams who have open, explicit methods of team decision making and conflict resolution are much more effective and efficient with their teamwork. For example, a team may decide to use only consultative decision making, allowing the chief engineer to be the ultimate decision maker on all recommended corrective actions. The team's role then becomes one of consultant to the chief engineer. If the team is unanimous in the absolution of decision making to the chief engineer, and the chief engineer is accepting of the responsibility of the decision, this may be a very effective decision-making method. In a different situation, the team may choose consensus as their decision-making method. The team may be comprised of equally concerned and authorized individuals, maybe even with competing intraorganizational goals. These teams may see the need to be committed to taking whatever time is necessary to achieve legitimate consensus on all decisions made. This may result in a lot of long and arduous days. But if each member of the team is committed to achieving consensus, and the SRFMEA is of such criticality to warrant the extra time and energy involved, this may be the best choice for a decision-making method.

Ultimately, every SRFMEA should be completed by a team of people. These people will spend a lot of time together. The concern of all stakeholders should be represented in this team, and all needs of expertise should be represented in this team. Then the team should spend some time on their teamwork processes, with decision making and conflict resolution being two critically important teamwork processes. Other teamwork processes in need of effective methods may be meeting management, personal accountability, knowledge management, and listening skills. The organization should work to provide teams with training and coaching on these teamwork processes as needed.

5.6.3   OPTIMISM OR PESSIMISM?

Any effective risk assessment runs a fine line between optimism and pessimism. An overly optimistic risk assessment may miss critical opportunities for risk mitigation. An overly pessimistic risk assessment may cause the team or organization to be frozen in its tracks, stuck with too much to do. Ideally the SRFMEA process uses two techniques to achieve the right balance of optimism and pessimism. The first technique includes having the right people on the team. The second technique involves how risks are identified and included in the SRFMEA.

Having the right people on the team for the SRFMEA process is critically important. All stakeholder interests should be represented on the team. This does not necessarily mean that the team includes the local community day care center director, but it may include the organization's human resource representative for employee benefits. Everyone on the team should be aware of the role they play as representing the stakeholders identified. And, all areas of expertise should be represented on the team. The team should be comprised of people with deep process and product knowledge for the functions being analyzed.

But for the correct balance of optimism and pessimism, more than just stakeholder representation and functional expertise is needed. The team should be comprised of a diversity of personality styles. We have all had those acquaintances for whom the glass is always half empty, and those others for whom the glass is always half full. A team made up of only optimistic personalities will miss important risks; an SRFMEA team made up of only pessimistic personalities may get paralyzed by trying to address too many improbable risks. The functional manager organizing an SRFMEA effort would be wise to choose a diversity of personalities for the team. Our experience has been that a team only slightly weighted toward a pessimistic personality tendency, with the appropriate expertise and stakeholder interest, is ideal.

The balance of optimism and pessimism can be further addressed through the SRFMEA process. After the right team has been generated—one that is adequately pessimistic to reach for deep risks, but not overly pessimistic to get stuck in the improbable—the way in which after-event reviews are conducted and operational definitions are created is an additional method to ensure balance. After each step of the process, especially the FMEA and RA prioritizations, it is recommended that the team immediately feed back criticism of their own process. For example, perhaps an SRFMEA team has set aside Thursday afternoon to complete their responsibility analysis of four of the SR categories. The team should allow time, at the end of the day, to generate critical feedback of their teamwork process experienced that day. Research has shown that teams that engage in after-event analysis achieve performance improvement. Allowing time for each team member to state what he or she felt went well with the teamwork processes during the Thursday event, and what he or she felt the team struggled with, will provide learning for improved performance for the next teamwork event. Building in a team self-assessment on the balance toward optimism/pessimism during these after-event analyses will allow for mid-task corrections to this teamwork element.

Operational definitions can also be important toward building a balance of optimism and pessimism to the SRFMEA teamwork. Careful definitions of severity class and occurrence class, and critical thinking toward the need to modify these definitions throughout the process, can help this balance. Organization-wide definitions and examples for severity class and occurrence class can provide consistency and optimism/pessimism balance across the organization. The organization should consider a knowledge management system, or library, of examples of severity ratings and qualitative occurrence ratings. The SRFMEA team could then rely upon these organization-level ratings to break through any overly optimistic or overly pessimistic tendencies.

5.6.4 TOO LITTLE, TOO LATE

Cautions for the SRFMEA process also include the quality of the corrective actions and the timing of the completion of the SRFMEA. An organization that does not promote a careful, thoughtful approach to the use of the SRFMEA process for the mitigation of risk may see the SRFMEA process completed as just a mindless task to be ticked off the "to do" list. Organizational culture, driven by the appropriate rewards, should be built for attention to an SRFMEA product of the highest quality. An artifact of the SRFMEA as task, not careful process, is the SRFMEA completed too late in the product design or function deployment. An after-the-fact SRFMEA is a symptom of an organization that does not see the SRFMEA process as a legitimate tool for the continuous improvement of social responsibility.

Mindless SRFMEAs are often completed for the wrong reasons and the wrong goal. We will call this observation of completing the SRFMEA just to get the task accomplished gaming the system. For example, one way in which an SRFMEA process can be gamed is for the team to achieve an SRFMEA result with inordinately high RPNs, only to be seen as heroes later when thoughtless or easy actions result in "improved" RPNs. Another game we have experienced is the listing of already planned actions tacked on to the SRFMEA. The team may be aware of a new piece of processing equipment to be purchased next year. Instead of digging deep to find corrective actions beyond the impact of this machine, they simply add on this already planned action to their SRFMEA. An organization that is overly focused on rewarding the improvement of RPNs, not careful, mindful risk assessment, may experience these games.

Another example of gaming the SRFMEA process is to load up the corrective actions with easy, not important, corrective actions. And finally, another SRFMEA failure may happen when the team is convinced of the solutions desired by the stakeholder without actually talking to, or involving, the stakeholder in the solution generation. An organization that rewards its SRFMEA teams by the number of corrective actions completed, without regard to the impact or quality of the actions, may experience these games.

Another caution for the SRFMEA process is the timing of the completion of the analysis during the design of the product or the deployment of the process. The goal of the SRFMEA process is the mitigation of social responsibility risk, not the remediation of damage already done. The earlier in the deployment, the more cost-effective mitigation can be, and the less risk of stakeholder harm. It has been our experience that organizations that do not engage in a concerted risk assessment during their product design or function deployment end up with an SRFMEA process completed in ways, or ending in results, that are not mindful of the goal of risk mitigation. And assuredly, those SRFMEAs done after the product is designed or the functional process is deployed will sit on a shelf as a "dead" document.

5.6.5 SRFMEA as Dead Document

For an organization that approaches the SRFMEA process as a way to improve its social responsibility, the SRFMEA and the RA are viewed as temporary outcomes, or just phases, of a much larger process. For these organizations, concertedly engaged in the mitigation of social risk, the identification of stakeholders, the teamwork nature of the SRFMEA process, and the detailed process analysis—the dynamic, living, day-to-day actions happening in the organization—are just as important as the final artifacts of the spreadsheets. However, for those organizations that choose to give cursory service to social responsibility, and may see the completion of the SRFMEA process as a task to accomplish in order to pass an audit, the SRFMEA and the RA will be dead documents sitting in an electronic library, or a printout filed in someone's desk. Often these are documents that are rarely used, seldom looked at, and never modified. Our advice to organizations that are using the documents only as evidence of their social responsibility program is to just stop applying resources to the creation of the documents. If the documents are not to be used in a living process, they are of no use. This organization could save effort and resources by concluding its program altogether.

When the SRFMEA and RA documents are living documents, there will be clues that indicate the organization's wisdom on the nature of the whole SRFMEA process. There will be evidence of a regular review of the documents. The information in the analyses will be used in project management. SRFMEA corrective actions will appear in project tracking management systems. They will be posted for many to see. These risk analyses will be used as foundations for strategies and business decision making. SRFMEA will be a process desired and well utilized, not something to be feared or endured.

5.7 CONCLUSION

As your organization embarks on the use of SRFMEA as a process of continuous improvement for social responsibility, be aware of our cautionary advice. Approach the SRFMEA as a dynamic, living process, not as a set of spreadsheets. Use a team approach and be careful of who is on the team in terms of stakeholder representation, expertise, and personality. And finally, beware of organizational rewards or leadership emphasis that may cause the process to generate the wrong results. Avoiding the creation of an organization that approaches SRFMEA as just a task to be accomplished and creating organizations that reward careful, thoughtful social risk mitigation is our goal.

Social responsibility is an ideal. It is a target of continuous improvement. It is a business imperative. Improving social responsibility performance requires a process-based approach of improvement. The time is right to gain the ability to significantly improve the socially responsible behavior of our companies, schools, governments, and other organizations through the definitions and guidelines being built by the International Organization for Standardization. And if we recognize the similarities of the social responsibility movement to the quality movement, we may be able to accelerate the achievement of our goals through the use of well-worn quality tools. However, we need action, not just talk.

The possibility of members of our organizations behaving irresponsibly is a risk. The failure mode effects and analysis is a robust tool, used by many professions, toward the effort to mitigate risk. We hope that we have shown you how to take this well-worn tool and apply it to the new opportunity of improving socially responsible behavior. We have taken an emerging international guideline on social responsibility, ISO 26000, our experience as practitioners of quality improvement tools, and our knowledge on organizational behavior and the management of risk to develop this tactical tool. Our hope is that you may now take this tool and apply it as a solution to move beyond the discussion of the need for socially responsible behavior to the ability to just do it.

6 A Case Study in Social Responsibility Performance Improvement

Nothing is more difficult than to introduce a new order. Because the innovator has for enemies all those who have done well under the old conditions and lukewarm defenders in those who may do well under the new.

—**Niccolo Machiavelli**

The following chapter demonstrates a fictional case study about an organization struggling with the desire to become more socially responsible. This type of organizational effort is always noble, and usually difficult. Doing the right thing is never easy. Organizations faced with making difficult choices, choices that change culture, climate, and behavior, need to understand their overall objective, and have tools at the ready to facilitate these difficult decisions. Following the whole seven-step process of the SRFMEA can help. Let's peek in on a fictional company, SRPharma, and see how it uses SRFMEA as a tool to improve its social responsibility performance.

SRPharma is an up and coming pharmacy that has found success locating its services inside hospitals (see Figure 6.1 for some statistics on SRPharma). It opened its first pharmacy in 2004 inside a hospital with two locations in South Texas. Due to the influx of winter Texans, it was able to expand to two more hospitals (two locations each) in the third year of operation. As SRPharma looked to expand, it decided to focus in other southern areas of the country that have large transitional, retiree populations. In 2008 it opened more pharmacies, two hospitals in Arizona (two locations each) and one hospital in Florida (two locations). Upon starting the business CEO Jane Domino, CFO Jeff Risak, and COO Randy Parker, while not educated as pharmacists, felt strongly that they wanted to ensure all of their business practices were socially responsible. They believed the nature of their business was saving lives and alleviating suffering. They sought an ideal of operating in a manner of responsibility that would continuously expand this effort, such that the outcomes of all of SRPharma's actions would contribute to the alleviation of suffering, not just the dispensing of medication. They hired Sherry Sasso, part-time, as corporate affairs director to lead this new effort. Sherry had recently helped Maui-Wauwii Coffee Company improve its social responsibility performance. However, she is new to the SRFMEA process. Having recently attended an American Society for Quality

Company name:	SRPharma
Core business:	Pharmaceutical sales inside of hospitals
Company sales:	$2.5 million
Production:	100 million prescriptions annually
Number of employees:	240
Number of locations:	12 (6 different hospitals)
Geographic locations:	United States (Texas, Arizona, Florida)

FIGURE 6.1   SRPharma statistics.

(ASQ) conference, where this new tool was taught in a workshop, Sherry is anxious to take her previous experience, combined with her new skill, and apply it to SRPharma's goals. Sherry is the only one of the group that is aware of the impending struggle. She applauds Jane, Jeff, and Randy's desire, but she recognizes their naiveté in thinking that this effort will be easy.

They encouraged and supported their socially responsible stance through the creation of a mission, vision, and value statements for the pharmacy that stressed the importance of the safe distribution of pharmaceuticals. The potential for catastrophic failures causing serious injury or death to their customers is a real risk for SRPharma. A few years ago Randy had instituted a very strong quality control program. After witnessing a multi-million-dollar settlement of their primary competitor, Jack-mart, for deaths occurring from the inadvertent distribution of expired drugs, Jeff had demanded such a program. If this same thing happened at SRPharma it would wipe them out. Randy was very happy and confident of the work his team had accomplished in deploying this program of total quality management. He felt that embarking on a program of social responsibility might resemble this program of total quality management.

The SRPharma executive team also wanted to position their organization as being socially and environmentally responsible. When they saw a recent report on the news about three illegal aliens arrested in the process of making several deliveries to Jack-mart pharmacies, they again took their cue. They began to wonder how, as an organization, they could protect themselves from this type of oversight. Jane recognized that social and environmental irresponsibility was not just bad management, but also put the company at risk—hence her search for a corporate affairs director who would lead the team to mitigate this risk. However, in this tight economy she was given the approval for only part-time resources. Jane was excited that Sherry had agreed to join the team. The board of directors had agreed to Sherry's plan of using SRFMEA to locate, prioritize, and alleviate both the real and public relations risks. This was a trial of trust. Sherry had also mentioned a new ISO 26000 guideline for social responsibility that was on the horizon. Because of their recent work with their quality management program, and the achievement of ISO 9000 certification, Jane was confident that this was the right time and the right action for SRPharma. With the baby boomer demographics falling into their strategic plans for growth, the last thing Jane wanted was a story like Jack-mart's pulling down her success.

Sherry's first step was to pull together a cross-functional team of employees, pharmacists, and managers to assist in the identification of potential risks based on the ISO 26000 guidelines. One of their first steps in the process was to evaluate the pharmaceutical value chain to understand who were the key stakeholders of their business. Some research that was conducted identified the following key stakeholder groups:

- Patient (customer of the service)
- Physician (referral of patient for the service)
- Pharmacy (service provider)
- Hospital (referral of patient for the service)
- Health insurers (service providers)
- Employees (service providers)
- Government (regulatory)
- Pharmaceutical manufacturers (supplier of inventory)

The leadership team understood the importance of knowing all of the key stakeholders from projects and problems experienced in their prior work settings. Once Sherry was sure that the team understood who the key stakeholders were, she wanted to educate the team on the core subjects of social responsibility as identified by ISO 26000. These core subjects with underlying issues are

- Organizational governance
- Human rights
  - Due diligence
  - Human rights risk situations
  - Avoidance of complicity
  - Resolving grievances
  - Discrimination and vulnerable groups
  - Civil and political rights
  - Economic, social, and cultural rights
  - Fundamental rights at work
- Labor practices
  - Employment and employment relationships
  - Conditions at work and social protection
  - Social dialogue
  - Health and safety at work
  - Human development and training in the workplace
- The environment
  - Prevention of pollution
  - Sustainable resource use
  - Climate change mitigation and adaptation
  - Protection and restoration of the natural environment
- Fair operating practices
  - Anticorruption
  - Responsible political involvement
  - Fair competition

- Promoting social responsibility in the sphere of influence
- Respect for property rights
- Consumer issues
  - Fair marketing, information, and contractual practices
  - Protecting consumers' health and safety
  - Sustainable consumption
  - Consumer service, support, and dispute resolution
  - Consumer data protection and privacy
  - Access to essential services
  - Education and awareness
- Community involvement
  - Community involvement
  - Education and culture
  - Employment creation and skills development
  - Technology development
  - Wealth and income creation
  - Health
- Social investment

Once the leadership team had a foundational understanding of their stakeholders and ISO 26000, they began the work involved in the SRFMEA process (Figure 6.2).

STEP 1: IDENTIFYING A PROCESS OF STUDY THROUGH THE SIPOS

Choosing a process upon which to focus was tough. Jane wanted to focus on the company's hiring practices, because of the recent headlines from Jack-mart's hiring of illegal persons. Her main focus was to prevent slanderous news stories. Randy wanted to focus on the dispensing of medications. His recent work on the quality management system should make this work effortless. He had too much to worry about with the plans for opening up four new operating centers by the end of the year. The last thing he needed was disruption to these operating plans. Jeff wanted to focus on the control of inventory. There had been some recent rumblings about legal liability if narcotics went unaccounted for. And Jeff was never very trusting of the pharmacists. After all, what a great job for a drug dealer—access to anything you wanted. Although the last inventory cycle count went well, Jeff couldn't help but think maybe it went too well.

Sherry stepped in and reminded them that they would eventually need to tackle all of these social responsibility risks. They were only choosing the *first* process to attack. And since this would be a learning opportunity, and because of their previous work with improving their quality control systems in the dispensing of medication, and the potential risks associated with this critical process, the team chose to use this process. The first process of focus chosen was the dispensing of medication. They then began to follow the seven-step SRFMEA process, which is an integration of the ISO 26000 core subjects and the FMEA process.

With the leadership team now ready to sponsor the social responsibility performance improvement of the dispensing medication process, Sherry pulled a

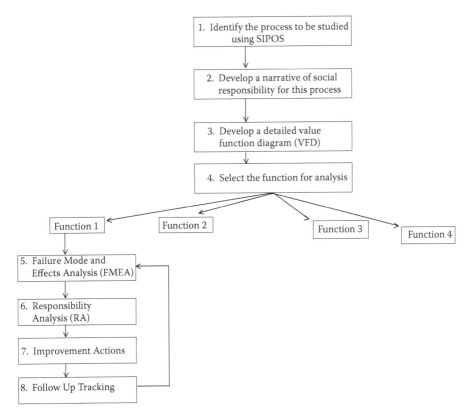

FIGURE 6.2   The SRFMEA process.

cross-functional team of pharmacists, employees, and managers together to create the SIPOS. In a single two-hour meeting this knowledgeable team was able to complete the SIPOS shown in Figure 6.3. Sherry facilitated the team in this brainstorming work by working backwards through S, O, P, I, and then S, as she had been shown in the ASQ workshop. The team had already identified their stakeholders. After making sure they hadn't forgotten anyone, this list went up on the board.

Then the team brainstormed all of the outputs that were delivered, through the dispensing of medicines, to these stakeholders. These outputs included both the product, the pills or liquids dispensed, and the information, such as the warning literature and the review of accuracy completed by the pharmacist.

Next the process itself was sketched, as shown in the value function part of the value/failure diagram (Figure 6.4). The whole process of consideration starts when the technician enters the prescription information, as noted from the doctor, into the computer system. The process of study will include the review of the labels from which medication will be dispensed, a visual review of the product itself, and the verification of the dispensed amount, as well as the pharmacist's review of accuracy. The studied process concludes with the medication, along with the warning and usage literature, placed in the bag and given to the customer.

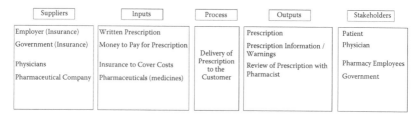

FIGURE 6.3   The SIPOS.

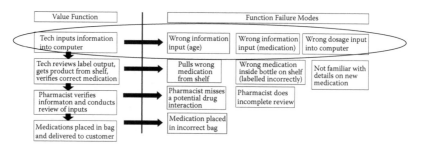

FIGURE 6.4   The value/failure diagram (VFD).

The team discussed including the payment of the medication in this process, but felt that this should be left out of the study. There is a concern for insurance fraud and identification fraud, but Sherry suggested that the team leave this work for future SRFMEAs. There was also a heated debate in the meeting about when the process of study should begin. SRPharma receives the prescription via a written paper from a doctor, an electronic prescription from those doctors in their network, or by phone from the doctor's office. There are differing levels of liability toward irresponsible behavior for each of these methods. For example, the technician could misread the doctor's handwriting on the piece of paper and cause great harm. Electronic systems are suspect to hacking and thus contributing to the trafficking of illegal narcotics. And the potential for fraud from phone-in prescriptions is a risk. Which of these, or none, or all, should be of consideration for the team?

Again, Sherry's advice was needed for the team to make progress. Since this is a learning process, Sherry advised the team to start the process of study with the process step immediately after the receipt of the prescription. Had this been a team already practiced in SRFMEA, Sherry would have advised the inclusion of the receipt of the prescription in the study. But Sherry was warned in the workshop that it is very easy to get overwhelmed in the completion of an SRFMEA. Many smaller risk assessments are better than a few complex ones.

After the hard work of gaining consensus on the boundaries of the process of study, the team brainstormed the inputs to the process. This includes any material, information, or energy needed to make the process function. As we can see from Figure 6.4, a lot of this input comes from the patient, or customer. SRPharma also needs to have the appropriate pharmaceuticals on the shelf. Lastly, in the completion of the SIPOS, the suppliers for each of these process inputs are brainstormed.

After this sometimes tense SIPOS completion session a few of the team members remarked to Sherry that they had not realized that the pharmacist was required to do data entry into the computer system for every prescription. Others were amazed at the relative simplicity of this critical process. And some members remarked that they had never thought of the prescribing physician as a stakeholder in the process. All in all, Sherry was very proud of the team's efforts. The goal of the SIPOS had been achieved: The cross-functional team had a much better understanding and appreciation of the process under study.

STEP 2: DEVELOPING A NARRATIVE

While Sherry was busy working with the cross-functional team on the SIPOS, she had given a list of questions to the executive team to answer. She wanted answers to questions like: Who are we as an organization? What is the goal of this analysis? Why do we want to improve social responsibility? How will this analysis be transparent to the rest of the organization? These questions could only be answered by the executives. After the SIPOS was done, she questioned the cross-functional team, too. She asked them questions like: How will you talk to your stakeholders? What is the planned timing of the analysis? How will best practices be found and communicated? How will you monitor and measure changes to process outcomes? The answers to these questions and others would become the information necessary to create a narrative. The narrative ensures that the goals and objectives of the risk assessment are met. The hopes and desires of the improvements to come from the corrective actions of the risk assessments should meet the strategic vision of the organization. Sherry took the answers to all of these questions, from the executives and the cross-functional team, and formulated the following short narrative:

> SRPharma is committed to the health of our customers. We recognize that the maintenance of health, recovery of health, and alleviation of suffering are why our customers seek our services. We choose to provide these services in a manner that preserves the social and ecological sustainability of the communities in which we operate.
>
> All employees of this organization are responsible, from the receipt of the drugs into the warehouse to the distribution clerk, for ensuring that all medication labeling matches the content of all containers. The traceability of pharmaceutical identification is critical to the dispensing of accurate products. If there are discrepancies, this needs to be brought to the pharmacist's attention immediately. Any incorrect dispensary of medication can cause irreparable harm to the customer.
>
> These critical consumer issues, as well as community involvement and labor practices, are important social responsibility subjects of the process of dispensing our products. We are actively committed to improving our performance by mitigating the risks associated with dispensing medication. We chose to improve our social responsibility behavior for our physicians, their patients, our customers, the members of our community, and our employees. Our goal is to minimize the potential for negative impact caused by unsustainable actions to these important stakeholders.

Sherry then showed the narrative to the executives. Jane requested a few editorial changes, specifically requesting the addition of the comment "If there are discrepancies, this needs to be brought to the pharmacist's attention immediately," as she had just been made aware of employee survey scores, which indicated a rift between some technicians and some pharmacists. She wanted to make sure that everyone understood that the pharmacist is ultimately liable, and thus responsible, in this regard. Sherry reminded Jane that it is the process, not the people, that cause errors in outcomes, but acquiesced on this small point. Then, Sherry showed the narrative to the cross-functional team. They were happy with the statement and ready to get to work on the SRFMEA.

STEP 3: DEVELOP A VALUE/FAILURE DIAGRAM

With the SIPOS complete, a brief value function analysis complete, and both the executive team and the cross-functional team satisfied with the SRFMEA narrative, the team had work ahead to focus the risk assessment. Sherry had learned in her workshop that in most processes there are steps within the process that are inherently more risk averse toward social and ecological failures than others. An easy way to highlight these process steps is through the value/failure diagram (VFD). The team had already finished listing a brief process flow, one that focused on the value being delivered at each process step, when they completed the SIPOS. Their work is shown on the left-hand side of the value/failure diagram shown in Figure 6.4. Sherry facilitated the team to brainstorm all of the possible failures for each of these process steps.

For example, the first process step involves the technician entering the prescription information into the computer. There are lots of things that can go wrong with this process step. The technicians demonstrated to the rest of the team that there were, on average, 15 fields of information that had to be typed into the computer at this process step—everything from the patient's age, to patient identifiers used to connect information of potentially harmful drug interactions, to information on the prescribing physician, to the medication and dosage. Because this information is manually typed into the computer, it is prone to error. And the risk of some of these failures could be life threatening.

Sherry facilitated the team's work to complete this type of analysis on each of the process steps. She wanted the team to recognize, and document, the most obvious potential failure modes. The value/failure diagram ends up looking a little like a stem-and-leaf diagram. The process steps with the most potential for failure have many items stacked up to the right of the process step. Since each process step is documented as a function, which adds value to the customer of the outcome of the process, these failures include not only catastrophic errors but also simply the failure to deliver value.

STEP 4: SELECTING THE FUNCTION FOR ANALYSIS

After Sherry's team was finished with the value/failure diagram the team recognized that there were many potential negative outcomes to the process they were analyzing. To choose just one value function upon which to focus an SRFMEA just didn't seem to make sense. How could they choose the wrong medication being placed in the bag as less critical than the wrong dosage entered into the computer? However, again, Shelly remembered the stern advice against overwhelming her team in their continuous improvement efforts. They weren't going to be able to eliminate all risks in a single SRFMEA; this is a continuous improvement initiative. She assured the team that they would be completing an SRFMEA on *all* of these value functions. The team was correct: Any one of these value functions was just too critical not to study. But, they needed to choose the *first* value function upon which the risk assessment would be completed.

The team decided, in this case, to start at the beginning. They felt that there might be improvements that could be made to the process step that involved the technician's data entry, which might remove some of the risk in subsequent process steps. So, in this case, rather than choosing the value function with the greatest potential for failure, they recognized that any failure in this whole process is unacceptable. And to conduct the risk assessment, they should break the task into the order of the process steps. Sherry was proud of her team's decisions. They would immediately deep-dive a risk assessment on the value function of the technician's information input in the process of dispensing medication.

STEP 5: FAILURE MODE AND EFFECTS ANALYSIS

Steps 1 through 4 had only taken the team three meetings and two weeks of information gathering. However, Sherry knew that the tough work would be coming in the next step. When completing the failure mode and effects analysis (FMEA) Sherry knew that the team would need to scrutinize all of the possible negative outcomes of their selected process and value function. They would go back to the seven social responsibility core subjects of ISO 26000 to gain guidance. They would need pessimists on the team, people who had seen things gone wrong. She knew that she was the consummate cheerleader, and needed to be careful of her own optimistic attitude.

The team used the ISO 26000 seven core subjects to identify all the possible failure modes of the value function of the technician entering the prescription information into the computer system. Sherry noted that it was very difficult to keep the team on task. They kept adding possible failure modes from the process step of the pharmacist's audit, or the verification of the product. Sherry had to work very hard to keep the team focused on only the value function of technician's data entry. In the end, they found potential failures in five of the seven core subjects. For example, they recognized that if the patient realized he or she had the wrong medication, and returned it to the pharmacy, by policy, that medication could not be dispensed again, and thus was added to the waste pharmaceuticals. If this happened frequently, it could negatively impact the environment by having all of these chemicals needlessly added to the hazardous waste stream. The team could not readily identify any impact to the core subject of community development. However, Marty, one of the technicians on the cross-functional team, had just read at iPharma.com that an unscrupulous pharmacy technician had shared knowledge about a local woman's contraceptive prescription to her mother-in-law. This was when the woman was presumably trying to get pregnant. This spiraled into serious consequences for the woman's marriage. Marty recognized that breaching privacy of the information the technicians' inputs can have negative community consequences.

Function	Technician inputs information	Technician into the computer	
Core Subject	**Social Responsibility Failure Modes and Causes**	**Failure Effects**	**Severity Class**
Organizational governance	No known impact		
Human rights	No known impact		
Labor practices	Process step without mistake proofing or audit creates extreme liability from technician errors	Stress and turnover at technician role	5
Environment	Wrong type of medication dispensed	Waste medication in hazardous waste stream	7
Fair operating practices	Underdosage of medication at full charge	Unfair charge to consumer or insurance company	3
Consumer issues	Wrong type of medication dispensed	Death, harm, or failure to heal patient	9
		Death or harm from drug interactions	9
	Wrong dosage of medication dispensed	Death, harm, or failure to heal patient	9
Social development	Technician breaches patient privacy	Patient embarrassment and legal liability	5

FIGURE 6.5   Failure mode and effects analysis.

After all of the potential failure modes and their effects were identified, Sherry facilitated the use of SRPharma's organizational definitions of severity class by the team (see Figure 6.5). Of course, physical harm to the customer is the worst possible outcome; these failure modes are given a severity class of 9. There was quite a bit of discussion on some of these ratings. For example, should stress experienced by the technician due to the criticality of the process step be rated more severe than the embarrassment by the customer from the breach of confidentiality? Consensus on some of these scores was difficult to achieve. Sherry reminded the team that only the highest severity scores will be addressed with corrective actions. So the difference between a severity score of 5 or a severity score of 7 is not significant when there are still severity scores of 9. This helped the team refocus and complete the scoring.

STEP 6: RESPONSIBILITY ANALYSIS

All of the failure modes identified on the FMEA were transferred to the responsibility analysis (RA) (Figure 6.6). The RA then used both qualitative and quantitative rationalization for the occurrence ratings. In fact, based upon the pharmacist's audit, SRPharma had created statistical process capability metrics for the error of incorrect medicine or incorrect dosage. This had been a big impact action during their quality

Function: Technician inputs information into the computer

Core Subject	Social Responsibility Failure Modes and Causes	Failure Effects	Severity Class	Occurrence Class	RPN	Remarks and Justification
Labor practices	Process step without mistake proofing or audit creates extreme liability from technician errors	Stress and turnover at technician role	7	7	77	Some employee complaints have arisen; turnover has been very high only for this employee group
Environment	Wrong type of medication dispensed	Waste medication in hazardous waste stream	7	3	73	Statistical probability of wrong medication = .0023
Fair operating practices	Underdosage of medication at full charge	Unfair charge to consumer or insurance company	3	5	35	Statistical probability of wrong dosage = .0146
Consumer issues	Wrong type of medication dispensed	Death, harm, or failure to heal patient	9	3	93	Statistical probability of wrong medication = .0023
		Death or harm from drug interactions	9	3	93	Statistical probability of wrong medication = .0023
	Wrong dosage of medication dispensed	Death, harm, or failure to heal patient	9	5	95	Statistical probability of wrong dosage = .0146
Social development	Technician breaches patient privacy	Patient embarrassment and legal liability	5	1	51	Technicians sign ethical statement; there is no known occurrence at SRPharma

FIGURE 6.6 The responsibility analysis.

management system deployment. Sherry had been taught to use the compound number method of responsibility priority number (RPN) at the SRFMEA workshop she attended. This resulted in the severity class 9 failure modes receiving the highest RPN. Sherry recognized that these same consumer safety issues, such as wrong medication and wrong dosage, were being actively addressed by SRPharma's quality control group. To prevent the quality improvement initiative and the social responsibility improvement initiative from stepping on each other's toes, she challenged this social responsibility team to dig a little deeper into the RPN prioritization.

The next highest RPN, with a score of 77, is the labor practice core subject issue of technician stress and employee turnover. Sherry encouraged the team to tackle this issue. When determining the occurrence class the team had been surprised to receive the employee turnover rates from human resources for the technicians. It was triple those of all other classes of employee. Clearly this was a problem. And the technicians on the cross-function team echoed this issue. The technicians recognized that they were the only SRPharma persons to see the physician's written prescription, electronic prescription, or hear the physician's voicemail. This meant that any misinterpretation on their part could cause catastrophic failure. Marty said this had caused plenty of sleepless nights for him.

And this RPN is immediately followed by a score of 75 for the environmental issue of incorrectly dispensed medicines being entered into the hazardous waste stream. This seemed like a great opportunity for improvement to the team. Sherry was very proud of the team's work on the RA. Neither of these issues, technician stress or waste medicines, had been a focus for SRPharma. The SRFMEA process had just uncovered opportunities for social responsibility performance improvement.

STEP 7: IMPROVEMENT ACTIONS

So far Sherry's cross-functional team had invested five meetings into the SRFMEA effort. However, the executive continuous improvement program monthly meeting was coming up next Thursday. She had been asked to present her team's progress. She really wanted the SRFMEA analysis work to be complete and to be ready to submit improvements for sponsorship at the executive meeting. She wanted Jane, Jeff, and Randy to see how rapidly improvement could be identified through the SRFMEA process. And, privately, she wanted them to recognize all the work that needed to be done to improve SRPharma's social responsibility performance; she wanted her part-time job to turn into a full-time job.

The team recognized that they would be at cross-purposes to the quality department if they chose to tackle the consumer issues. So they decided to inform the quality department of their analysis of the social responsibility failures of prescription errors and to offer help for any actions already identified by the quality department. However, on the labor practices and environmental issues, they alone needed to present corrective actions. For the technician stress issue, the team realized that the process already existed for the double-checking; they just didn't have the right information at the right place at the right time. The pharmacist's review of the dispensation included a review of all of the technician's work, except a review of the original data entry of the medicine and dosage. A very simple solution of adding

Function	Technician inputs information into the computer			
Core Subject	**Social Responsibility Failure Modes and Causes**	**Failure Effects**	**Remarks and Justification**	**Improvement Action**
Labor practices	Process step without mistake proofing or audit creates extreme liability from technician errors	Stress and turnover at technician role	Some employee complaints have arisen; turnover has been very high only for this employee group	Include review of original prescription (paper, electronic, or phone voice recording) at pharmacist's review and audit
Environment	Wrong type of medication dispensed	Waste medication in hazardous waste stream	Statistical probability of wrong medication = .0023	Investigate potential for use of returned medication for veterinary uses
Consumer issues	Wrong type of medication dispensed	Death, harm, or failure to heal patient	Statistical probability of wrong medication = .0023	See quality improvement program
		Death or harm from drug interactions	Statistical probability of wrong medication = .0023	See quality improvement program
	Wrong dosage of medication dispensed	Death, harm, or failure to heal patient	Statistical probability of wrong dosage = .0146	See quality improvement program

FIGURE 6.7   Improvement actions.

these factors to the pharmacist's review could be accomplished. The process change would require saving the original prescription, email, or voicemail from the doctor for direct audit by the pharmacist. The pharmacists were highly in favor of this addition. After all, all the other checks are negated if the technician's original interpretations of the prescription were wrong.

On the environmental issue, Marty, again using iPharma.com as a wealth of information, found that apothecary shops, which mix medicines for veterinarians, would take some pharmaceuticals, with traceability and expiration information secured, for reformulation. So, in those rare events, when the erroneous dispensing caused waste pharmaceuticals, this potential could be investigated. It wouldn't eliminate the waste stream, but it might reduce it. It was a first step in improvement, while the quality department team was rapidly working on eliminating the dispensing errors.

Sherry was ready to present these newfound opportunities and the team's recommendations to the executive team (Figure 6.7). Jane was very impressed that the team had already completed its analysis and was ready for recommendations. Randy was happy that the team recognized the need to defer the consumer safety issues to the quality department. Jeff, the CFO, was a little disappointed that none of the recommendations would result in an immediate cost savings. But Sherry explained that SRPharma would be paying less for hazardous waste disposal, and that there are significant hidden costs to the recruiting and training of new technicians. At that, he asked Sherry how fast she could finish 20 more SRFMEAs. Sherry replied, "Well, if I was working full time ..."

DISCUSSION QUESTIONS

1. If you were on the team and could override one of the three choices of core subject by the team, what core subject would you remove, what core subject would you add, and why?
2. Which stakeholder in the pharmaceutical value chain do you believe is the most important to this type of business and why?
3. What stakeholder(s) was left off the list? Why is this stakeholder important to the process selected? Are there any other elements missing from the SIPOS?
4. What more, if any, information is relevant and should be included in the narrative?
5. Would you have selected another value/failure combination? Why?
6. If you were a team member, what other core subjects do you believe are impacted by the selected function? What specific failure modes and causes could you identify? What would be the failure effect? How severe would this failure and effect be?
7. How sure are you that there is a remote chance of a wrong prescription being given to the customer? Would you recommend a qualitative or quantitative approach to rating the occurrence? Why?
8. What other actions would you suggest for improvements?
9. Discuss the team. Was anyone left out? What sources of team conflict should be anticipated? How should this be resolved?
10. What other actions would you suggest for improvement?

Glossary

CONCEPTS

Accountability: A recognition that no organization is perfect and a provision of trust in appropriate action when problems arise.

Actionists: Individuals within an organization who are responsible for taking action associated with any initiative, task, or element of work.

Actuarial risk assessment: The analysis of financial losses due to hazards or risks, primarily related to the insurance industry. It involves the assessment of the severity of loss and probability of loss due to financial catastrophe.

Aspects: A feature or facet of a system. In reference to social responsibility, it is an activity inherent to a system that may be at risk of irresponsible behavior.

Autism spectrum disorder: A class of psychological disorder typified by abnormal social interaction and communication.

Business impact analysis: Also known as business continuity planning. This is the response plan for a business to reconvene operations after a catastrophic event. Understanding the severity and probability of a disruption to business operations due to threats such as earthquake, fire, or flood.

Community involvement and development: As defined as a core subject in ISO 26000, community involvement and development concerns the active engagement of the organization in the concerns of the community in which the organization operates and the responsibility to create improvements in the community.

Consumer issues: As defined as a core subject in ISO 26000, consumer issues are the protective responsibilities to the consumers of the organization's goods or service.

Continuous improvement: A business management strategy that expects constant analysis and improvement through regimented change process.

Corporate philanthropy: The practice of corporate cash donations to charitable organizations.

Corporate social responsibility: See *social responsibility.* The application of social responsibility to the subgroup of organizations chartered as corporations.

Criticality analysis (CA): The part of the risk assessment when the failure modes are prioritized through the compilation of a risk priority number, as a combination of the severity and the occurrence of the failure mode.

CSR: See *corporate social responsibility.*

Detection: The ability to know that a failure is occurring as it occurs.

Environment: As defined as a core subject in ISO 26000, the environment considers the impact on the natural environment arising from the decisions and activities of the organization.

Ethical behavior: Acting with integrity, honesty, fairness, and concern for all stakeholders and the environment.

Failure effect: The consequences of a failure mode. It is what happens to the process or system when the failure mode occurs. For social responsibility this is the impact on society or ecology when irresponsible behavior happens.

Failure mode: The means by which the system fails. A description of the part and method of the system or process that breaks or causes the system or process to fail to function. For social responsibility, the failure mode is the breakdown of the system or process, which then creates a lack of ecological or social sustainability.

Failure mode effects and analysis: A specific type of risk assessment where failure modes are identified. The effects of these failure modes are also identified. The failure modes are ranked by their relative severity.

Fair operating practices: As defined as a core subject in ISO 26000, fair operating practices are those dealing with other organizations completed in an ethical manner.

FMEA: See *failure mode effects and analysis.*

Function: The purpose or utility of a process or system.

Genchi Gembutsu: A term used in the deployment of the Toyota Production System that indicates the need to personally witness events in order to understand them. This term is typically translated as "go and see."

HAACP: See *Hazard Analysis and Critical Control Point System.*

Hazard analysis: A type of risk analysis prevalent in industries concerned with worker and consumer safety. It involves the assessment of the severity of loss and probability of loss due to safety hazards.

Hazard Analysis and Critical Control Point System: A program of the mitigation of risks of food and pharmaceutical safety. It concerns the identification of process control aspects, or critical control points, identified through a hazard analysis.

Health impact assessment: An impact assessment methodology used to determine the potential severity and probability of effects on the health of a population.

Human rights: As defined as a core subject in ISO 26000, human rights are the basic rights to which all human beings are entitled because they are human beings.

Impacts: The influence, effect, or outcome of an action. With respect to social responsibility, these are the effects of behaviors on the social and ecological systems in the surrounding environment.

Kaizen: A term used in relation to the Toyota Production System to mean small, incremental, but continual process improvements to be completed by everyone in the organization.

Knowledge criticality analysis: A risk assessment performed on potential failure of the loss of knowledge critical to the ongoing concerns of the organization. It involves the assessment of the severity of loss and the probability of loss due to the failure of an organization to retain critical knowledge.

Labor practices: As defined as a core subject in ISO 26000, labor practices encompass all policies and practices relating to work performed within, by, or on behalf of the organization.

Lean Production: A colloquial term used to indicate the business management strategy deployed by Toyota Motor Corp. This involves a focus on the elimination of non-value-added activity and waste from processes.

Narrative: A written description. In reference to social responsibility, it is a description of the boundaries, assumptions, goals, and intentions toward a target of improvement.

NGO: See *nongovernmental organization.*

Nongovernmental organization (NGO): A social organization of treaty, guideline, or behavioral importance, but not related to any specific government or sovereignty.

Objectives: The achievement attained from a set of actions. For continuous improvement activities, the objective is the desired outcome of improvement. It is the reason for the improvement activity.

Occurrence class: The classification, through either rank order or rating, of the probability of loss due to system failure.

Organizational governance: As defined as a core subject in ISO 26000, organizational governance is the system by which an organization makes and implements decisions.

Plan-do-check-act: An iterative process of problem solving made popular by W. Edwards Deming. The problem solver plans to act upon the system in order to correct a problem, does the corrective action, checks for the outcome of the correction, and then acts upon the outcome to either plan additional correction or control the appropriate correction.

Process: An assembly of many actions in series and parallel, typified by inputs, outputs, and transforming activities.

Profound knowledge: W. Edwards Deming's concept of knowledge needed by managers of business and operational processes. It includes the appreciation of a system, knowledge of variation, theory of knowledge, and knowledge of psychology. It is only after the achievement of profound knowledge that a process should become the target of continuous improvement.

Quality movement: The period of time, from post-World War II to the 1980s, in which business management strategy moved from arts and crafts and command and inspection, to process control, statistical analysis, and process-oriented problem solving. It was this period of time when significant increases in manufactured product quality were achieved.

Quantitative risk analysis: A risk analysis that considers the severity and probability of failure. This analysis is typically completed on electrical or mechanical systems through probablistic/stochastic modeling.

Reasonably available control measures: A term typically used in the pollution control field, to indicate the practical limits of the mitigation or control of hazards.

Responsibility analysis (RA): For social responsibility, this is the part of the risk assessment when the failure modes are prioritized through the compilation of a risk priority number, as a combination of the severity and the occurrence of the failure to behave in a socially or ecologically sustainable way.

Reliability: The probability of a successful outcome. A measure of dependability.

Risk: An unknown, usually negative, future state. A chance of loss.

Risk assessment: A quantification of risk. This is a relative measure of two factors: the magnitude of the loss and the probability of the occurrence of the risk. For social responsibility, this is the analysis of the severity of not behaving in a socially or ecologically sustainable way combined with the analysis of the probability of the irresponsible behavior through an understanding of the organizational system being analyzed.

Risk management: The actions and systems deployed to attempt to mitigate the potential for system loss or failure.

Risk priority number (RPN): The compilation of a ranking that combines severity and occurrence within a criticality analysis. This can be accomplished either by multiplying the severity rating and the occurrence rating or by creating a compound number where the digit in the tens place is the severity rating and the digit in the ones place is the occurrence rating.

Root cause problem solving: The process of uncovering the fundamental conditions and actions that lead to problematic outcomes.

RPN: See *risk priority number.*

Severity class: The classification, through either rank order or rating, of the magnitude of potential loss experienced by system failure.

SIPOS: See *suppliers-inputs-process-outputs-stakeholders.*

Six Sigma: A business management strategy that targets the reduction of process variation and the achievement of the elimination of defects.

Social responsibility (SR): The responsibility of an organization for the impacts of its decisions and activities on society and the environment, through transparent and ethical behavior that contributes to sustainable development, health, and the welfare of society; which takes into account the expectations of stakeholders, is in compliance with applicable law and consistent with international norms of behavior, and is integrated throughout the organization and practiced in its relationships. This includes products, services, and processes. Relationships refer to an organization's activities within its sphere of influence (from ISO 26000).

Social responsibility failure mode effects and analysis (SRFMEA): The use of failure mode effects and analysis as a risk assessment tool toward the risk of socially irresponsible behavior. It is a seven-step process that includes identifying the process to study, developing a narrative of social responsibility about the process, developing a value/function diagram, selecting the function(s) for detailed analysis, analyzing the failure modes and effects of the function, conducting a responsibility analysis, and identifying and tracking improvement actions.

SR: See *social responsibility.*

SRFMEA: See *social responsibility failure mode effects and analysis.*

Stakeholder: A person or group that can be affected by the organization's activities and decisions.

Suppliers-inputs-process-outputs-stakeholders (SIPOS): A graphical technique of listing all the suppliers, inputs, outputs, and stakeholders of a process. It also includes a summarized, graphical depiction of the process.

Sustainability: The ability of social and ecological systems to endure over very long periods of time. Sustainable systems do not exhaust their required inputs without the reuse or recycle of system output back to input.

Targets: The object of a goal. For continuous improvement activities, the target is the social or ecological issue or system for which improvement is expected.

Threat assessment: An assessment conducted on the risks to public safety, such as violence or terrorism.

Total quality: See *Total Quality Management.*

Total Quality Management (TQM): A business management strategy that sets the expectation for everyone in the organization to be aware and responsible for the quality of work delivered to the downstream internal or external customer.

Toyota Production System (TPS): A business management strategy characterized by the relentless pursuit of the elimination of waste and non-value-added activities throughout the organization.

Transparency: The act of conducting business decisions in a nonsecret environment. Transparency involves open access by stakeholders to the information and process involved in organizational decision making.

STANDARDS AND GUIDELINES

INTERNATIONAL ORGANIZATION FOR STANDARDIZATION (ISO)

ISO 9000, as a series of standards for quality management systems
   ISO 9001:2008, *Quality Management Systems—Requirements*
   ISO 9004:2000, *Quality Management Systems—Guidelines for Performance Improvement*

ISO 14000, *Environmental Management*
ISO 26000, *Guidance on Social Responsibility*
TS 16949, *Quality Management Systems—Particular Requirements for the Application of ISO 9001:2008 for Automotive Production and Relevant Service Part Organizations*

QUALITY MANAGEMENT INDUSTRY STANDARDS

AS 9100, *The European Association of Aeropsace Industries*
QS 9000, *The Automotive Industry Action Group*
TL 9000, *QuEST Forum for the International Telecommunications Industry*

GOVERNMENT STANDARDS

United States, Sarbanes–Oxley Act of 2002, Public Company Accounting Reform and Investor Protection Act of 2002, a.k.a. SOX

U.S. Green Building Council, Leadership in Energy and Environmental Design, a.k.a. LEED

U.S. Department of Defense, MIL-STD-1629, *Procedure for Performing a Failure Mode, Effects and Criticality Analysis*, a.k.a. FMEA and CA, or FMECA

NGO STANDARDS AND GUIDELINES

Software Engineering Institute at Carnegie Mellon University, Capability Maturity Model Integration, a.k.a. CMMI

World Health Organization, Good Manufacturing Practices, a.k.a. GMP or cGMP

United Nations Environment Program, Global Reporting Initiative, a.k.a. GRI or G3

World Commission on Environment and Development, *Our Common Future*, a.k.a. sustainable development

PEOPLE

Crosby, Philip B. (1926–2001). Credited with the quality concept of zero defects.

Deming, William Edwards (1900–1993). Credited as the initiator of the quality movement in the United States and the creator of the concept of profound knowledge of management.

Feigenbaum, Armand V. (1922–). Credited with the creation of Total Quality Management.

Friedman, Milton (1912–2006). A Nobel Prize-winning economist.

Ishikawa, Kaoru (1915–1989). Credited with the creation of the cause-and-effect diagram.

Juran, Joseph M. (1904–2008). Credited with many quality management tools, such as the popularity of the Pareto principle, and cross-functional management principles that contributed to the quality movement and Total Quality Management.

Ohno, Taiichi (1912–1990). Credited as the creator of the Toyota Production System.

Shewhart, Walter Andrew (1891–1967). Credited as the originator of statistical process control.

Shingo, Shigeo (1909–1990). Credited with bringing the Toyota Production System to the attention of U.S. businesses in the 1980s.

Taguchi, Genichi (1924–). Credited with the Taguchi loss function, which recognized that any deviation from optimal conditions results in a loss to society.

References and Resources

Automotive Industry Action Group. (1995). *Potential failure mode and effects analysis (FMEA): Reference manual.* Author.

Aven, T. (2003). *Foundations in risk analysis: A knowledge and decision-oriented perspective.* West Sussex, England: John Wiley & Sons, Ltd.

Becker-Olsen, K. L., Cudmore, A., & Hille, R. P. (2006). The impact of perceived corporate social responsibility on consumer behavior. *Journal of Business Research, 59,* 46–53.

Bollier, D. (1996). *Aiming higher: 25 stories of how companies prosper by combining sound management and social vision.* New York: AMACOM.

Bomann-Larsen, L., & Wiggen, O. (Eds.). (2004). *Responsibility in world business: Managing harmful side effects of corporate activity.* Tokyo: United Nations Press.

Canwest News Service. (2007, August 15). *Parents have trouble finding toys not made in China.* Retrieved June 26, 2009, from http://www.canada.com/topics/news/national/story.html?id=a56528bf-e88b-41ee-81cd-2b3fa3a8beec&k=39042

Carrefour Oman launches "Small Change Big Difference" campaign. (2007, July 7). Retrieved May 24, 2009, from www.newswiretoday.com/news/20693/

CNN. (2009, March 18). *Smelly drywall ruining homes, owner says.* Retrieved from http://www.cnn.com/2009/US/03/18/chinese.drywall/index.html?eref=rss_topstories

CNN most accountable global 500 firms. Retrieved June 26, 2009, from http://money.cnn.com/galleries/2008/fortune/0811/gallery.accountability.fortune/index.html

Coles, G., Fuller, B., Nordquist, K., & Kongslie, A. (2005). Using failure modes analysis and criticality analysis for high-risk processes at three community hospitals. *Joint Commission on Accreditation of Healthcare Organizations, 31,* 132–140.

Council on Economic Priorities. (1998). *The corporate report card.* New York: Penguin Group.

Cutler, A. (2009). *Illegal aliens delivering drugs to Wal-Mart.* Retrieved June 28, 2009, from http://www.abc4.com/content/news/top%20stories/story/Illegal-aliens-delivering-drugs-to-Wal-Mart/iw6AW1k610W_iiRXgsPEcQ.cspx

Ethisphere. (2007). *2007 world's most ethical companies.* Retrieved June 26, 2009, from http://ethisphere.com/wme-2007-rankings/

Friedman, M. (1970, September 13). The social responsibility of business is to increase profits. *New York Times Magazine.*

Frynas, J. G., & Pegg, S. (2003). *Transnational corporations and human rights.* Houndsmills, UK: Palgrave MacMillan.

Global Reporting Initiative. Retrieved from http://www.globalreporting.org/

Handfield, R. B. & McCormack, K. (Eds.). (2008). *Supply chain risk management: Minimizing disruptions in global sourcing.* New York: Auerbach Publications.

Harrison, J., & Freeman, R. (1999). Stakeholders, social responsibility, and performance: Empirical evidence and theoretical perspectives. *Academy of Management Journal, 42,* 479–485.

Harvard Business School. (2008). *Lessons learned: Straight talk from the world's top business leaders—Going green.* Boston: Harvard Business School Publishing.

Hollender, J., & Fenichell, S. (2004). *What matters most: How a small group of pioneers is teaching social responsibility to big business, and why big business is listening.* New York: Basic Books.

Jennings, M. M. (1996). *Case studies in business ethics.* Minneapolis, MN: West Publishing.

Mayard, Y. M. (2007). *Consumers' and leaders' perspectives: Corporate social responsibility as a source of a firm's competitive advantages.* Doctoral dissertation, University of Phoenix.

Newton, L., & Schmidt, D. P. (2004). *Wake up calls: Classic cases in business ethics.* Toronto: Thomson South-Western.

Paper vs. plastic—The shopping bag debate. Retrieved May 25, 2009, from www.blog. greenfeet.com/index.php/paper-vs-plastic-the-shopping-bag-debate/reducing-your-footprint/12

Passas, N., & Goodwin, N. (Eds.). (2007). *It's legal but it ain't right: Harmful social consequences of legal industries.* Ann Arbor: University of Michigan Press.

Pitta, D. A., & Laric, M. V. (2004). Value chains in healthcare. *Journal of Consumer Marketing, 21,* 451–464.

Sloss, D. (2006, January 1). Do international norms influence state behavior? *George Washington International Law Review.*

Estrella, C. A. (2008, June 29). *Small changes—Big difference: Kanu Hawaii asks folks to make a commitment.* Retrieved May 24, 2009, from www.honoluluadvertiser.com/apps/pbcs.dll/article?AID=2008806290322

Social Responsibility Newsletter (2007, March). Issue 8. Retrieved April 7, 2009, from www. iso.org

Staib, R. (2005). *Environmental management and decision making for business.* Houndmills, UK: Palgrave MacMillan.

Top ten environmental tips. Retrieved May 24, 2009, from www.kab.org.au/01_cms/details. asp?ID=100

USA Today. (2007, September 3). *Child labor use rises in China.* Retrieved from http://www. usatoday.com/news/world/2007-09-03-china-labor_N.htm

USA Today. (2007, October 29). *India activists decry gap child labor.* Retrieved from http:// www.usatoday.com/money/industries/retail/2007-10-29-gap-child-labor_N.htm

Webb, D. J., Mohr, L. A., & Harris, K. E. (2008). A re-examination of socially responsible consumption and its measurement. *Journal of Business Research, 61,* 91–98.

Wilson, R. & Crouch, E. A. C. (2001). *Risk-benefit analysis.* Boston: Harvard University Press.

Appendix A: Failure Mode Effects and Analysis Blank Form

APPENDIX A: Failure Mode Effects and Analysis Blank Form

Function			
Core Subject	**Social Responsibility Failure Modes and Causes**	**Failure Effects**	**Severity Class**
Fair operating practices			
Labor practices			
Human rights			
Environment			
Organizational governance			
Consumer issues			
Social development			

Appendix B: Responsibility Analysis Blank Form

APPENDIX B: Responsibility Analysis Blank Form

Core Subject	Social Responsibility Failure Modes and Causes	Failure Effects	Severity Class	Occurrence Probability	Responsibility Priority Number	Remarks and Justifications
Fair operating practices						
Labor practices						
Human rights						
Environment						
Organizational governance						
Consumer issues						
Social development						

Function

Appendix C: SRFMEA Process Flow

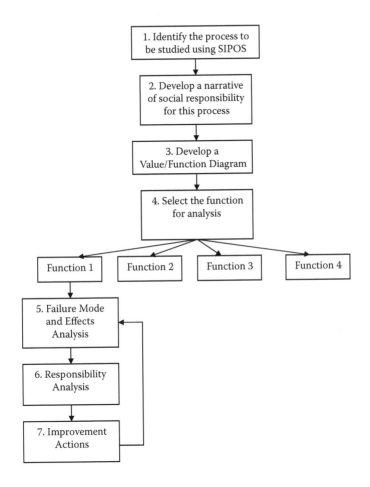

Appendix D: SRFMEA Completion Checklist

ORGANIZATIONAL SUPPORT

- Process selected is critical to organizational success
- Process selected supports strategic initiatives
- Leaders of the organization recognize this process as important for social responsibility
- Leaders of the organization support the SRFMEA process
- There is a manager/mentor that is the champion of the SRFMEA process
- Adequate resources have been assigned to complete the SRFMEA
- Appropriate cross-knowledge team members have been selected to complete the SRFMEA
- A systematic process for review of SRFMEA elements is in place and valued by the leadership of the organization

COMPLETION OF SIPOS

- Developed exhaustive list of stakeholders
- Identified industry-specific stakeholders
- Developed exhaustive list of outputs of the selected process that are important to the stakeholders
- Developed exhaustive list of inputs to the selected process
- Identified suppliers of each of the inputs identified
- Completed high-level process flow to bound the project

DEVELOPMENT OF NARRATIVE

- Completed narrative elements in paragraph format
- Documented the profile of the organization relative to the project selected and its purported benefit to the organization
- Documented the goal(s) of the SRFMEA analysis and why it is important to the success of the organization
- Documented the context of why this analysis is important to the organization and what socially responsible benefits the organization hopes to gain
- Documented who are the key stakeholders and why they are important, also explained why any stakeholders were excluded

- Documented all of the potential risks associated with the selected project/ process, the impact of the risks to specific stakeholder groups, and how those risks, if mitigated/eliminated, benefit the organization
- Documented how the analysis process will be conducted, which resources will be involved, what are the expectations of the analysis process, and what are the expected results
- After completion of the SRFMEA, returned to the narrative to document a summary of the results from the analysis; discussed how risk mitigation/ elimination will benefit the organization and the key stakeholders

COMPLETION OF VALUE FUNCTION ANALYSIS

- Considered all of the detailed process steps for the selected process
- Combined the detailed steps into groupings that result in added value for the stakeholder
- Identified an exhaustive list of failures for each value function

SELECTION OF VALUE FUNCTION(S)

- Selected value function with the greatest social responsibility risk
- Selected value function with the greatest potential for failure
- Selected value function that will potentially impact organizational performance
- Selected value function that supports the strategic initiatives of the organization

COMPLETION OF THE FAILURE MODE
AND EFFECTS ANALYSIS (FMEA)

- All seven ISO 26000 social responsibility core subjects are documented on the FMEA
- Identified an exhaustive list of failure modes for each of the seven ISO 26000 core subjects
- Identified an exhaustive list of failure effects for each of the seven ISO 26000 core subjects
- Developed a team methodology for agreement on the severity class of the failure
- Team agreement on the severity class of each failure mode
- Transferred all seven ISO 26000 core subjects, failure modes, failure effects, and severity rating to the responsibility analysis document

COMPLETION OF THE RESPONSIBILITY ANALYSIS (RA)

- Identified the likelihood of occurrence for each of the items identified in the FMEA step
- Developed a team methodology for agreement on the occurrence class of the failure

- Team agreement on the occurrence class of each failure mode
- Selected the appropriate risk priority number (RPN) approach for your organization/industry (product approach or compound approach)
- Calculated the RPN number
- Prioritized the results based on the RPN
- Selected a few critical items to improve based on organizational impact
- Transferred the selected SRFMEA items to improve to the action item document

DEVELOPMENT OF IMPROVEMENT ACTIONS

- Identified current RPN value and have mechanism on the form for documenting new value after improvement completed
- Identified risks and justification for each item on the improvement action and tracking report
- Identified improvement action for each item transferred to the improvement action and tracking report
- Identified person responsible for each action item
- Identified date when each action item will be completed
- Identified new RPN value after completion of the action

Appendix E: SRFMEA
Precautions Checklist

- Don't be too pessimistic or too optimistic when brainstorming potential failures for each of the core ISO 26000 SR subjects.
- Focus on the SRFMEA process—completion of all of the steps—not on the SRFMEA document.
- Utilize all seven ISO 26000 core subjects (fair operating practices, labor practices, human rights, environment, organizational governance, consumer issues, and social development) in the failure analysis and every improvement iteration.
- Ensure that all high RPN SR elements have improvement actions—do not overlook any improvements with high RPN values.
- Completion of the SRFMEA is done most effectively with a cross-functional team—if you find yourself completing alone, stop.
- Complete the SRFMEA as part of the existing design, system, and process reviews—don't wait until all the decisions have already been made.
- Complete the entire process on each iteration of continuous improvement—be aware of technological, political, and regulatory changes that might affect your analysis in new ways.
- Be aware of new information about the core ISO 26000 subjects that might impact your failure modes or ratings in the SRFMEA process.
- Be conscious of industry-specific social responsibility concerns and incorporate into the SRFMEA process.
- Standardize and adjust the severity class and occurrence class based on the needs of your product, service, and industry.
- Complete an SRFMEA for those functions identified in the value function flow analysis that tie to your organizational strategic objectives.
- Be exhaustive in identifying potential failures and effects for each of the seven ISO 26000 core subjects.

Appendix F: Team Interview Questions

- Has a cross-functional team been indentified that represents all key players impacting the selected process?
- Have all of the key stakeholders been identified?
- Have outputs for each stakeholder been identified?
- Have all of the inputs to the selected process/project been identified?
- Have suppliers for each of the inputs to the selected processes/projects been identified?
- Has the narrative about the process been completed to include goals, definitions, key concerns, etc.?
- Have all of the key value functions been identified and detailed in the process flow?
- Have failures been identified for each of the value functions in the detailed process flow?
- Do all team members agree on the function(s) to focus on for the SRFMEA?
- Have you identified failure modes for all seven of the core social responsibility elements?
- For each failure mode, have you identified the potential effects of the failures?
- Is there team agreement on the elements of the severity table?
- Is there team agreement of the elements of the occurrence table?
- Has the team decided to use a qualitative or quantitative approach?
- Has the team decided to use a compound RPN or a multiplier RPN?
- Based on the selected actions, does the team have members that are responsible for implementing the necessary actions?
- Has the team made formal arrangements to review progress and follow up on the action items?
- Has the team taken responsibility for incorporating new failures, effects, actions, changes, etc., associated with the original SRFMEA?

Appendix G: Typical Failure Modes for ISO 26000 Core Subjects

ORGANIZATIONAL GOVERNANCE

- Lack of organizational policy on the key principles of accountability, transparency, ethical behavior, respect for stakeholders, respect for the rules of law
- No systems for tracking and/or reporting on social and environmental outcomes in addition to financial outcomes
- No active participation and involvement with key stakeholders
- Lack of promotion of under-represented groups to senior level positions
- Lack of use of resources efficiently (financial, human, and natural)
- Lack of systems to track and ensure that elements of risk are addressed and resolved

HUMAN RIGHTS

- Lack of clear, consistent message about the importance of human rights within the organization
- Lack of consideration of differences with regards to human rights in all of the countries in which a company operates
- Lack of systems to track and ensure that human rights issues are addressed and resolved
- Lack of a methodology (SRFMEA, for example) for assessing human rights risks
- Lack of guidance policies to those organizations directly linked, such as suppliers and customers
- Lack of processes for resolving grievances
- Discrimination against a protected or underrepresented group

LABOR PRACTICES

- Conditions of work do not comply with local, state, national, or international laws
- Appropriate conditions of work are not applied (work hours, holidays, time off, etc.)
- Wages are not paid directly to the person(s) doing the work
- Employees not compensated for overtime work

- Equal pay is not provided for equal work
- Conditions of work conflict with local social and religious customers and traditions
- Employer does not provide and ensure that the work environment is safe
- Lack of training and development for the employees
- Lack of policies that ensure open communication with regard to social issues

THE ENVIRONMENT

- Lack of system for tracking pollution created by the organization
- Lack of identification of business actions and decisions that impact the environment
- Lack of action taken to resolve environmental issues that impact the community at large
- Lack of policies and practices for reuse, recycle, and remanufacture of products/services utilized in the business process
- Lack of identification and action taken to mitigate impact on climate change
- Lack of identification and action associated with protecting the natural environment

FAIR OPERATING PRACTICES

- Lack of identification of risk associated with corruption
- Lack of systems and structure for reporting incidents of corruption
- Lack of awareness and inaction associated with competition legislations
- Lack of procedures/processes for educating and communicating the importance of social responsibility to suppliers, contractors, customers, and those actively involved in the execution of your business activities
- Lack of policy on ensuring, investigating, and acting on issues associated with property rights

CONSUMER ISSUES

- Lack of policies, procedures, and processes that ensure fair marketing practices
- Lack of risk assessment of the impact of the product/service failure on the consumer's health and safety
- Lack of policies, procedures, and processes on communicating, educating, and acting on products/services that negatively impact the ecosystem
- Lack of systems to support active communication and resolution of issues from the consumer
- Lack of efficient systems to protect the information and privacy of consumers
- Lack of effective programs for educating and increasing awareness of the products/services with regard to the consumer

COMMUNITY INVOLVEMENT AND DEVELOPMENT

- Lack of action involving participation in the local community
- Lack of involvement in the local community to encourage and promote education
- Lack of active effort to provide employment and skill development in the local community
- Lack of active effort to increase transfer and increase technological knowledge of the local community
- Lack of involvement in economically stimulating the local environment and ensuring that the disadvantaged benefit from income generation
- Lack of involvement in the community to increase awareness, reduce incidences, and improve overall health of the citizens of the community
- Lack of involvement, investment, and participation in social projects within the community

Appendix H: SRFMEA Example—Supply Chain

Example: Supply Chain

Step 1. Identify the process to be studied using SIPOS

Suppliers	Inputs	Process	Outputs	Stakeholders
V.P. purchasing Supplier quality New potential suppliers	Supplier due dilligence Supplier quotes Supplier product samples Supplier capacity study	Selection of suppliers for new products	Certified supplier Certified components from new suppliers	Employees External customer Consumer Management team

Example: Supply Chain

Step 2. Develop a narrative for social responsibility
In paragraph form, as a team, answer the following questions while creating a narrative of thoughts, goals, and patterns which are relevant to the analysis

Profile	As an organization we are committed to ensuring that all entities that supply us with products and services are not actively participating in conduct that would be considered socially irresponsible.
Goal	The goal of this analysis is to identify any potential risks that may exist in the process of selecting new suppliers. The assessment of this risk will ensure that we continue to foster support for our economic, social, andenvironmental stewardship.
Context	We are interested in ensuring that all new suppliers in the process of conducting business on our behalf do notprovide products that will endanger the environment or the community at large.
Stakeholders	The key stakeholders are the employees of the organization who must make decisions in the selection process on which supplierbest meets all the product, process, and social responsibility criteria. Other stakeholders include the external customer and theultimate consumer who could be negatively impacted by a socially irresponsible supplier.
Risks	The potential risks that the company wants to avoid are related to the selection of a new supplier that participates, condones, or is blind to socially irresponsible behavior.
Process of Analysis	The process of selecting the supplier will involve not only the validation of the supplier's product quality, process efficiency, butalso whether they are socially responsible citizens.
Summary of Analysis	A summary of the SRFMEA analysis suggests that there are catastrophic failures that could cause sickness, injury, or death thatneed to be addressed not only with new suppliers, be also reassessed with existing suppliers.

Example: Supply Chain

Step 3. Develop a Value/Failure Diagram. Identify the failure modes for
each value function in the process flow. Select only one function at a time for analysis.

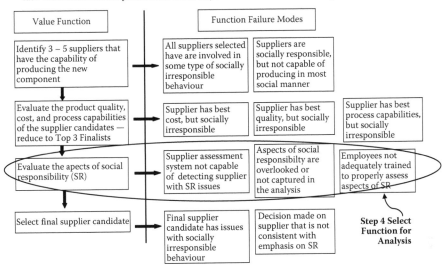

Example: Supply Chain

Step 5. Social Responsbility Failure Mode and Effect Analysis

Function: Projects Executed and Results Reported			
Responsibility Risk	**Social Responsibility Failure Modes**	**Failure Effects**	**Severity Class**
Fair Operating Practices	Supplier is not conducting business in a manner that supports fair competition	Supply disruptions impact the organization	5
Labor Practices	Supplier is hiring underage workers	Children hired are not attending school which is affecting the local environment	7
Human Rights	Supplier is not paying workers for overtime	Possible employee uprising that could cause production / supply delays	6
Environment	Supplier is not correctly disposing of waste per international requirements	Waste is polluting the local community	7
Organizational Governance	Supplier does not have a system for identifying and reporting the amount of waste disposed	No formal knowledge of the type and amount of waste being disposed and how it impacts the community	4
Consumer Issues	Supplier is using illegal / dangerous chemicals to produce its products	Consumer is injured or dies as a result of use of the product; production/ supply delays	9
Community Involvement and Development	Lack of systems for identifying and reporting waste has resulted in waste disposal that has polluted the water and affected the local community	Members of the community become sick or die as a result of the waste disposal problem	9

Example: Supply Chain

Step 6. Responsibility Analysis

Function: Projects Executed and Results Reported

Responsibility Risk	Social Responsibility Failure Modes	Failure Effects	Severity Class	Occurrence Probability	RPN	Remarks and Justification
Consumer Issues	Supplier is using illegal / dangerous chemicals to produce its products	Consumer is injured or dies as a result of use of the product; production/supply delays	9	7	97	Need a system for assessing whether supplier product does not use dangerous chemicals
Community Involvement and Development	Lack of systems for identifying and reporting waste has resulted in waste disposal that has polluted the water and affected the local community	Members of the community become sick or die as a result of the waste disposal problem	9	7	97	Need to develop a system to identify and measure impact of waste on society

Example: Supply Chain

Step 7. Follow-Up Tracking

Function: Projects Executed and Results Reported

Responsibility Risk	Social Responsibility Failure Modes and Causes	Failure Effects	Remarks and Justification	Improvement Plan
Consumer Issues (97)	Supplier is using illegal / dangerous chemicals to produce its products	Consumer is injured or dies as a result of use of the product; production/ supply delays	Need a system for assessing whether supplier product does not use dangerous chemicals	1. Develop assessment tool for assessing all suppliers; 2. Assess top suppliers to ensure that illegal / dangerous chemicals are not being used; 3. Go back and assess all suppliers to ensure due diligence with regards to consumer issues
	Who: R. Moore	**When:** 12/2/2008		**New RPN** 63
Community Involvement and Develop-ment (97)	Lack of systems for identifying and reporting waste has resulted in waste disposal that has polluted the water and affected the local community	Members of the community become sick or die as a result of the waste disposal problem	Need to develop a system to identify and measure impact of waste on society	1. Develop a system for identifying all of the waste streams at supplier locations; 2. Evaluate new suppliers for their potential waste streams associated with your product; 3. Verify whether claims have been made against the supplier with regards to pollution in the community
	Who: H. Duckworth	**When:** 12/13/2008		**New RPN** 65

Appendix I: SRFMEA Example—Household Cleaning Improvement

SRFMEA is not only for corporations; it can also be utilized for efforts to become more socially responsible in your personal life. This example demonstrates the potential social responsibility risks that can be associated with cleaning your house.

1. Identify the process to be studied using SIPOS.

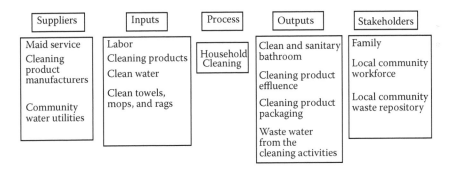

2. Develop a narrative for social responsibility.

We, the Johnson family, desire to provide clean and sanitary bathrooms in the family's home by using social responsibility principles. The whole family—Jason, Marie, Tiffany, and Jacob—will be involved in making decisions about how we will be conducting our household cleaning. We want to make sure that the decisions we make result in a healthy home and a healthy community. We recognize that even this small decision by this small family can have an impact on our society. If we choose to use illegal labor or toxic products, we are contributing to a long-term problem with sustainability. If we choose to use legal labor and safe products, we may send a message to our neighbors to also make better decisions.

3. Develop a value/failure diagram. Identify the failure modes for each value function in the process flow. Select only one function at a time for further responsibility analysis.

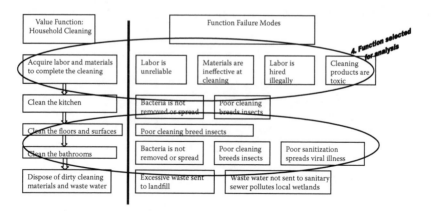

5. Failure mode and effects analysis:

Function	Acquire labor and materials to clean the bathrooms		
Core Subject	**Social Responsibility Failure Modes and Causes**	**Failure Effects**	**Severity Class**
Organizational governance	No known impact		
Human rights	Illegal employment	Human trafficking	9
		Loss of taxes	3
Labor practices	Unequal share of labor within the family	Gender inequality	3
		Domestic strife	1
		Unfair control or exclusion to decisions	1
	Unfair pay of hired labor	Unfair impact (up or down) on local house cleaning labor costs	1
Environment	Use of toxic cleaning products	Pollution of community water source	9
		Promotion of pollution by product manufacturer	3
		Pollution of community landfill disposal area	9
		Harmful exposure of family members and laborers to toxins	9
Fair operating practices	No known impact		
Consumer issues	Supplier cleans too frequently	Waste of products and money	3
	Supplier cleans too infrequently	Unsanitary conditions	3
	Poor quality of cleaning service	Unsanitary conditions	3
Social development	No known impact		

5A. Failure mode effects severity table:

		Severity Class		
		1	3	9
Organizational governance	Will this failure affect governance (reporting) structure?	No effect of family decision making	Some negative impact to family harmony	Source of strife in family relationships
Human rights	Will this failure cause humans' freedom?	No effect on liberty of others	Long-term negative impact on human rights	Immediate harmful human rights impact
Labor practices	Will this failure negatively impact fair labor?	No negative effects on employment relationships	Some unfair labor practices	High risk of labor exploitation or unfair practices
Environment	Will this failure have an impact on the natural environment?	No toxic effluence	The creation of some toxic effluence	Known toxins created with environmental impact
Fair operating practices	Will this failure have an effect on the industry at large?	No industry impact	Long-term harmful impact on the home cleaning industry	Immediate harmful fair operating impact
Consumer issues	Will this failure cause safety, health, or damage to the consumer?	No impact to the consumer	Long-term harmful impact on consumers	Immediate harmful impact on consumers
Social development	Will this failure have an impact on society at large?	No community impact	Long-term harmful impact on the community	Immediate harmful impact on the community

6. Responsibility analysis:

Function	Acquire labor and materials to clean the bathrooms					
Core Subject	Social Responsibility Failure Modes and Causes	Failure Effects	Severity Class	Occurrence Class	Responsibility Priority Number	Remarks and Justification
Human rights	Illegal employment	Human trafficking	9	1	91	Not probable, but very bad when it happens
		Loss of taxes	3	9	39	If illegal employee is paid "under the table," it is highly probable that taxes will not be collected
Labor practices	Unequal share of labor within the family	Gender inequality	3	3	33	Concern of teaching equal household responsibilities to our son and daughter
		Domestic strife	1	1	11	We will work it out
		Unfair control or exclusion to decisions	1	3	13	We will work it out
	Unfair pay of hired labor	Unfair impact (up or down) on local house cleaning labor costs	1	9	19	If we pay too much, our neighbors may also have to; same if we pay too little
Environment	Use of toxic cleaning products	Pollution of community water source	9	9	99	The rinse water from any toxic chemicals used goes right into our city sewer system
		Promotion of pollution by product manufacturer	3	9	39	If we buy it, they will make it
		Pollution of community landfill disposal area	9	3	93	Paper towels and empty bottles may have some impact on the landfill, albeit diluted
		Harmful exposure of family members and laborers to toxins	9	3	93	Skin and air exposure to chemicals in use or spilled
Consumer issues	Supplier cleans too frequently	Waste of products and money	3	9	39	Timing of service needs to be carefully planned
	Supplier cleans too infrequently	Unsanitary conditions	3	3	33	Timing of service needs to be carefully planned
	Poor quality of cleaning service	Unsanitary conditions	3	9	39	Inspection of service quality needs to be carefully planned

6A. Failure mode effects occurrence table:
Qualitative Analysis

Occurrence	Definition	Occurrence Probability Level
Frequent	A high probability of occurrence. Probable failure during most process cycles.	9
Occasional	An occasional probability of occurrence. Some likelihood of a failure during at least one process cycle.	3
Extremely unlikely	The probability of failure is almost zero.	1

7. Follow up tracking:

Function	Machinery provided for office use: copiers, fax, computers, cell phones, PDAs, desk phones			
Core Subject	**Social Responsibility Failure Modes and Causes**	**Failure Effects**	**Remarks and Justifications**	**Improvement Plans**
Human rights	Illegal employment	Human trafficking	Not probable, but very bad when it happens	1. Educate the family on legal hiring laws 2. Ensure any labor hired has appropriate work visas 3. Complete the proper employer paperwork for income taxes and other fees for laborers
Environment	Use of toxic cleaning products	Pollution of community water source	The rinse water from any toxic chemicals used goes right into our city sewer system	1. Educate the family on the most environmentally nontoxic cleaning product choices 2. Make our own nontoxic cleaning products from natural materials 3. Reuse all spray bottles and containers 4. Use cloth cleaning towels to minimize landfill waste
		Pollution of community landfill disposal area	Paper towels and empty bottles may have some impact on the landfill, albeit diluted	
		Harmful exposure of family members and laborers to toxins	Skin and air exposure to chemicals in use or spilled	

Appendix J: SRFMEA Example—Initiative Implementation

Example: Initiative implementation (Six Sigma)

Step 1. Identify the process to be studied using SIPOS

Suppliers	Inputs	Process	Outputs	Stakeholders
Top Management	Top Management Support	Implement Six Sigma	Completed Projects	Internal Customers
Managers	Full-Time Black Belts		Cost Reductions Variation	External Customers
Six Sigma Steering Committee	Full-Time Master Black Belts Part-Time Green Belts Project Pipeline Program Goals Project Goals		Reduction Improved Customer Satisfaction	Management Team

Example: Initiative Implementation (Six Sigma)

Step 2. Develop a narrative for social responsibility.

In paragraph form, as a team, answer the following questions while creating a narrative of thoughts, goals, and patterns which are relevant to the analysis.

Profile	As an organization we believe that one route to success is to identify and reduce variation in all of our production and support processes. The team leading the implementation of this initiative is headed by the V.P. of Quality and all of the Quality Managers in each plant. This analysis is scheduled to be completed prior to the official kickoff in January 2010. This analysis will be transparent to the company due to its integration into the primary responsibilities of the quality department. This initiative and analysis has support from the CEO and top leaders of the firm. They have allocated time for each team member to work on this analysis on a daily basis until complete. They also support the dedicated resources of Black Belts and Master Black Belts.
Goal	The goal of this analysis is to identify any potential risks that exist that have the potential to derail this initiative. During the implementation and integration of this initiative we will take action that will equally support our economic, social, and environmental stewardship.
Context	We are interested in using this initiative to improve the quality of life for our employees through more effective work practices. We see this initiative as a way to improve the satisfaction of our customers and also positively affect our key stakeholders. This initiative should identify projects that ultimately result in reduced organizational costs which should satisfy all stakeholders. The analysis could potentially impact policy regarding how to manage displaced labor as a result of reductions.
Stakeholders	The key stakeholders are the leaders of the organization, the suppliers, shareholders, employees, and the customers. All stakeholders will ultimately benefit from having more effective processes with reduced variation. The key stakeholders will receive a presentation by the executive leader in charge of the implementation. This presentation will offer a rationale for the initiative and any expected changes to the organizatinal culture. Performance of the implementation and results of the project will be made available for all stakeholders to view electronically. The overall performance of the initiative will be included in the business update sessions at all levels within the organization.
Risks	The potentially negative outcome from this initiative will be the identification of ways to reduce headcount with more effective processes. The shareholders and customers both benefit from the outcomes in terms of more profit and better customer service, respectively. The awareness of risk will be identified using the SRFMEA process and the communication will occur via an online communication medium. Best practices will be shared using the same online medium. A severe failure will occur if employees are laid off as a result of a project or reduction in variation. The probability of displaced labor as a result of the implementation of this initiative is relatively high.
Process of Analysis	The process of implementation and integration will require adequate levels of human resources (Black Belts, Master Black Belts, and Green Belts). The outcomes will require resources to be accepting of the recommended changes to processes and potentially organizational culture.
Summary of Analysis	Summary of the analysis shows that each Six Sigma project must include analysis of all seven SR elements to ensure that all added costs and additional waste is accounted for and/or integrated into the total project savings.

Step 3. Develop a value/failure diagram. Identify the failure modes for each value function in the process flow. Select only one function at a time for analysis.

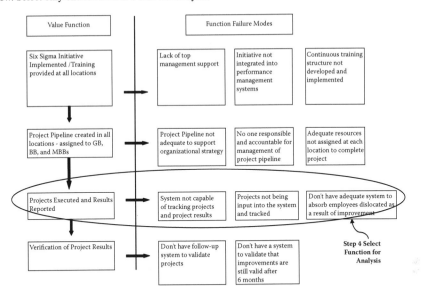

Example: Initiative Implementation (Six Sigma)
Step 5. Social Responsbility Failure Mode and Effect Analysis

Function: Projects Executed and Results Reported			
Responsibility Risk	**Social Responsibility Failure Modes**	**Failure Effects**	**Severity Class**
Fair Operating Practices	No known impact		
Labor Practices	Determining which person gets absorbed and how the absorption process will take place; policy that no one will get laid off as result of project improvement	Person relocated based on unfair labor criteria and/or practice	5
Human Rights	No known impact		
Environment	Project implementation results in additional waste that must be processed	More waste than expected to be disposed of	5
Organizational Governance	No known impact		
Consumer Issues	Change to process affects product quality as delivered to the customer or ultimate consumer	Quality defects in the field as a result of change related to a project	2
Community Involvement and Development	No known impact		

Example: Initiative Implementation (Six Sigma)
Step 6. Responsibility Analysis

Function: Projects Executed and Results Reported

Responsibility Risk	Social Responsibility Failure Modes	Failure Effects	Severity Class	Occurrence Probability	RPN	Remarks and Justification
Labor Practices	Determining which person gets absorbed and how the absorption process will take place	Person relocated based on unfair labor criteria and / or practice	5	9	59	No policy in place to absorb employeer whose jobs were eliminated
Environment	Project implementation results in additional waste that must be processed	More waste than expected to be disposed of	5	5	55	Need evaluation of waste as result of project deducted from project results
Consumer Issues	Change to process affects product quality as delivered to the customer or ultimate consumer	Quality defects in the field as a result of change related to a project	2	3	23	Need to include evaluation of change in control plan for the product

Example: Initiative Implementation (Six Sigma)
Step 7. Follow-Up Tracking

Function: Projects Executed and Results Reported

Responsibility Risk	Social Responsibility Failure Modes and Causes	Failure Effects	Remarks & Justification	Improvement Plan	
Labor Practices (59)	Determining which person gets absorbed and how the absorption process will take place	Person relocated based on unfair labor criteria	No policy in place to absorb employees whose jobs were eliminated	1. Develop global policy that no one will be eliminated as a result of an improvement project. 2. Develop a process. for absorbing displaced employees as a result of an improvement project.	
	Who: R. Moore	**When:** 11/6/2008		**New RPN**	30
Environment (55)	Project implementation results in additional waste that must be processed	More waste than expected to be disposed of	Need evaluation of waste as result of project	1. Include in the project analysis an environmental impact study. 2. Account for any disposal required and its impact on the environment. 3. Discount any costs associated with the waste from the project savings.	
	Who: H. Duckworth	**When:** 11/30/2008		**New RPN**	15
Consumer Issues (23)	Change to process affects product quality as delivered to the customer or ultimate consumer	Quality defects in the field as a result of change related to a project	Need to include evaluation of change in control plan for the product	1. Identify any potential quality issues in the control plan. 2. Ensure that item is added and evaluated in the PFMEA.	
	Who: R. Moore	**When:** 12/1/2008		**New RPN**	6

Appendix K: SRFMEA Example—Manufacturing

Example: Manufacturing (Product Assembly)

Step 1. Identify the process to be studied using SIPOS

Suppliers	Inputs	Process	Outputs	Stakeholders
Component Supplier	Raw Materials	Assemble Product	Finished Goods	Internal Customers
"Sub assembly" Supplier	"Sub assembly" Materials		Scrap/Waste	External Customers
Maintenance Department	Equipment		Output Reports	Management Team
H.R. Department	Documentation		Inventory	NGOs
Quality Department	Employees			Shareholders
Engineering Department				Employees
				Consumer

Example: Manufacturing (Product Assembly)

Step 2. Develop a narrative for social responsibility
In paragraph form, as a team, answer the following questions while creating a narrative of thoughts, goals,
and patterns which are relevant to the analysis

Profile	As an organization we believe that our strategy for success is to provide the highest quality product to a wide mix of customers. The quality of our product is managed in concert with our other key strategic focus areas of low cost and global appeal. This analysis will be undertaken with the interest in ensuring that we consider the impact of our product assembly on all key stakeholders and that we become more socially responsible citizens.
Goal	The goal of this analysis is to identify the potential social responsibility risks associated with the production of our products.
Context	The context of this analysis is to understand in more detail the impact that our product has on the various stakeholders and to determine whether there are social responsibility risks that need to be addressed associated with the production process is the boundary of focus for this analysis.
Stakeholders	The key stakeholders are the management team of the organization, the suppliers, consumers, shareholders, employees, NGOs, and the customers both internal and external. Evaluating the impact of our production process from the perspective of all stakeholders is of utmost importance in this analysis.
Risks	The risk associated with evaluating the social responsibility risks with regards to all stakeholders will open up the organization to knowledge about risks that we are not able to address in the short term.
Process of Analysis	The outcomes from the process of analysis will be important. The identifed failure modes for each stakeholder for each of the seven core elements of social responsibility. The process will focus on assessing the social responsibility risk of all key stakeholders with regards to our production process.
Summary of Analysis	Concerns with regards to packaging have the potential to affect the organization internally (waste/costs) and externally (supplier noncompliance and/or consumer displeasure). This is an area of the organization that we need to address across all product lines.

Step 3. Develop a Value/Failure Diagram. Identify the failure modes for each value function in the process flow. Select only one function at a time for analysis.

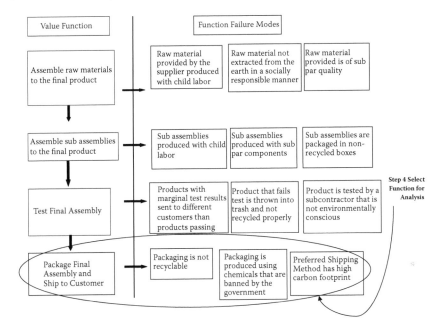

Example: Manufacturing (Product Assembly)
Step 5. Social Responsbility Failure Mode and Effect Analysis

Function: Package Final Assembly and Ship to Customer

Responsibility Risk	Social Responsibility Failure Modes	Failure Effects	Severity Class
Fair Operating Practices	No known impact		
Labor Practices	Packaging supplier is cited for using child labor	Loss of business due to affiliation with supplier: Interruption in service due to stop shipments from supplier.	8
Human Rights	No known impact		
Environment	Packaging used for finished goods is not recyclable	Generating waste and unnecessary costs associated with repurchase of packaging material.	6
Organizational Governance	No known impact		
Consumer Issues	Packaging is not recyclable. Product is not recyclable.	Consumer purchases competitor product because they have better product reuse/recycle program.	6
Community Involvement and Development	No known impact		

Example: Manufacturing (product assembly)
Step 6: Responsibility Analysis

Function: Projects executed and results reported

Responsibility Risk	Social Responsibility Failure Modes	Failure Effects	Severity Class	Occurrence Probability	RPN	Remarks and Justification
Labor practices	Packaging supplier is cited for using child labor	Loss of business due to affiliation with supplier; interruption in service due to stop shipments from supplier	8	7	87	Policy in place to assess suppliers to ensure that they follow appropriate practices
Environment	Packaging used for finished goods is not recyclable	Generating waste and unnecessary costs associated with repurchase of packaging material	6	9	69	Need evaluation of waste generated and associated costs
Consumer issues	Packaging is not recyclable; product is not recyclable	Consumer purchases competitor product because it has better product reuse/recycle program	6	9	69	Need to evaluate competitor packaging and reverse engineer product for analysis

Example: Manufacturing (product assembly)
Step 7: Follow-Up Tracking

Function: Projects executed and results reported

Responsibility Risk	Social Responsibility Failure Modes and Causes	Failure Effects	Remarks and Justification	Improvement Plan
Labor practices (87)	Packaging supplier is cited for using child labor	Loss of business due to affiliation with supplier; interruption in service due to stop shipments from supplier	Policy in place to assess suppliers to ensure that they follow appropriate practices	1. Ensure that assessment practice is in place in all locations and is conducted correctly
	Who: R. Moore	**When:** 7/6/2009		**New RPN: 35**
Environment (69)	Packaging used for finished goods is not recyclable	Generating waste and unnecessary costs associated with repurchase of packaging material	Need evaluation of waste generated and associated costs	1. Evaluate waste generation from packaging 2. Identify costs associated with repurchase of packaging materials 3. Implement waste improvement plan
	Who: H. Duckworth	**When:** 9/30/2009		**New RPN: 33**
Consumer issues (69)	Packaging is not recyclable; product is not recyclable	Consumer purchases competitor product because it has better product reuse/recycle program	Need to evaluate competitor packaging and reverse engineer product for analysis	1. Evaluate competitor packaging and see if is recyclable 2. Evaluate competitor product to see if has reusable/recyclable components 3. Evaluate competitors to determine if they have a closed-loop process for returns, recycle, reuse, remanufacture, etc. 4. Implement closed-loop program as identified or needed
	Who: R. Moore	**When:** 11/1/2009		**New RPN: 47**

Index